25 Years of Reporting Down-to-Earth Benefits

$PINOFF

97

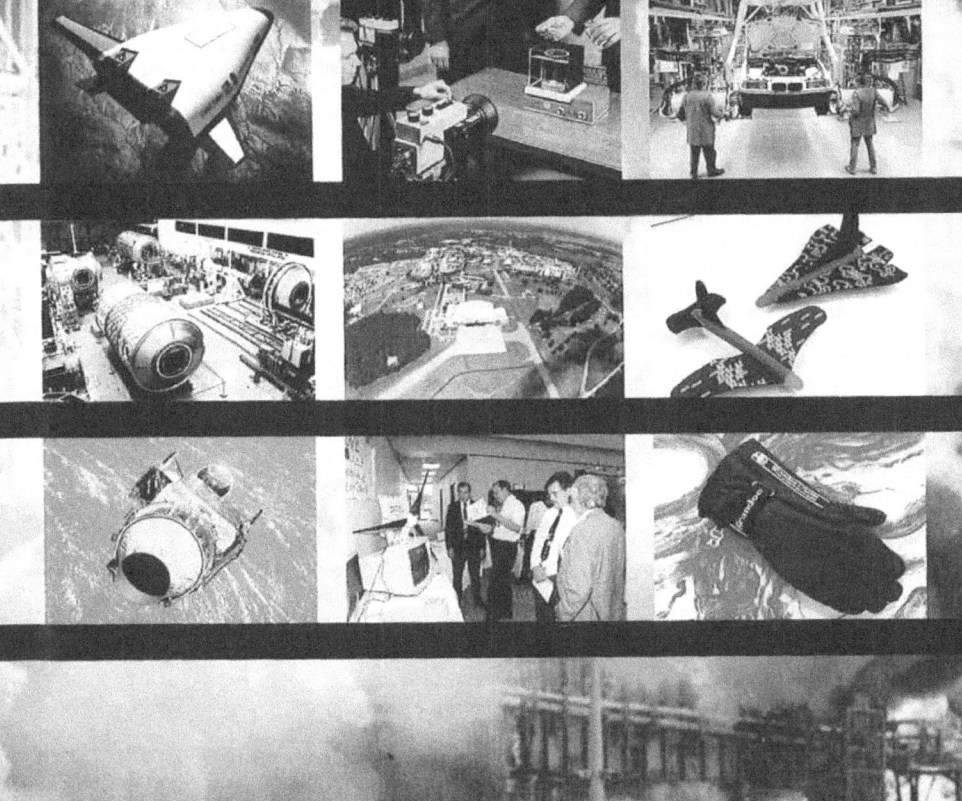

National Aeronautics and
Space Administration

Foreword

For a quarter of a century we have been writing about the effect NASA research and development has on you, the individual taxpayer, on industry and its commercial successes, and on the quality of life for humankind. In no small measure, this organization's scientific and engineering endeavors have favorably impacted our existence on Earth.

Past, current, and future aerospace programs will continue to provide technological advances that keep our industry in the forefront of global competition, contribute to major improvements in the health field, and improve lives in general.

Federal agencies are experiencing large-scale funding cuts in an effort to decrease the national deficit. NASA's Strategic Plan, as well as its reorganization, realignment, and restructuring, have positioned the Agency to continue its mandate: the expansion of frontiers and the exploration of that which lies beyond the horizon—space, planets, and other solar systems. In aeronautics, NASA will facilitate major improvements in global civil aviation impacting safety, revolutionary technology leaps cutting the development cycle for aircraft, and access to space resulting in reduction of payload cost to low-Earth orbit.

At the same time, a new way of doing business ensures that the technologies developed will have maximum commercial potential. This will be accomplished through proactive involvement of the private sector from the onset.

With the help of the greatest scientific and engineering minds of this great country, NASA will push forward and continue its research. Resulting transferrable technologies will find their way into products and services benefitting every aspect of life—manufacturing, jobs, health, income, convenience, exports, sales, and recreation.

Our efforts in aerospace exploration are supported by the expertise of nine NASA field centers and the Jet Propulsion Laboratory in disciplines such as robotics, environment, life sciences, materials, sciences and engineering, bioengineering, microgravity research, thermal efficiencies, satellite technologies, and wind tunnel testing. The centers also ensure that the technologies they develop quickly reach U.S. companies. They assist industry in remaining globally competitive as well as enriching the quality of life for all.

A sampling of research and development activities carried out at the field centers during the past 25 years is presented with great pride in this publication. And it is their technologies that enable NASA to contribute so successfully to the development and enhancement of products and services, to increased U.S. productivity, and to enhancing life in general in the areas of medicine, environment, entertainment, recreation, and education.

All spinoffs cited in this publication in previous years and those introduced here today are a tribute to the human spirit that brings forth technological progress. It honors entrepreneurs and visionaries in industry who have a keen eye for commercial success, improvements in health, and a cleaner environment. They display a never-ending search for technological advancement and excellence.

Daniel S. Goldin

Administrator, National Aeronautics and Space Administration

Introduction

Since 1973, this publication has provided you with examples of products and services developed as a direct result of the transfer of NASA developed technologies to private industry. Then—we featured fire retardant materials; now—we talk about non-invasive cardiac monitors. Then—we reported about air pollution detection devices; now—we discuss sensors for environmental control. Then—we heard about vitreous carbon materials to be used for prosthetics; now—we design major propulsion systems used in the development of energy saving jet engines.

Research and development in such areas as micro-circuitry, fiber optics, enhanced imaging, material processing—to name just a few—has generated technology for decades. It is available to private industry in a vast storehouse easily accessible through NASA's Commercial Technology Network.

The application of NASA technologies by the private sector increases productivity through the development of new products and processes that meet consumer demands; in turn, this benefits the national economy, industrial efficiency and human welfare. At the same time, it helps meet international challenges and aid U.S. industry to stay globally one step ahead of every scientific and technological innovation.

The return in benefits (spinoffs) to you the taxpayer, through new industries, new products and services, and improved quality of life, represents a substantial dividend on the national investment in aerospace research.

This publication will continue to heighten your awareness of the technology, know-how and assistance available for transfer, and the public benefit inherent in its utilization.

Spinoff 1997 is organized in three sections:

Section 1 summarizes the research and development efforts of the ten NASA field centers during the past 25 years; resulting technologies have in the past and present contributed to product commercialization, and will in the future affect economic and global decisions by industry.

Section 2 details the mechanisms NASA has utilized to transfer its technology and the types of assistance it renders to industry to advance technology transfer and commercialization efforts (for contacts see NASA's Commercial Technology Network).

Section 3, the focal point of this publication, features stories representative of success by manufacturers and entrepreneurs in developing commercial products and services that improve the economy and life for all humankind.

I hope you find NASA's latest spinoffs as interesting and informative as I do.

Robert L. Norwood

Dr. Robert L. Norwood

Director, Commercial Technology Division
National Aeronautics and Space Administration

Contents

Foreword 3

Introduction 5

Aerospace Research and Development

NASA Headquarters and Centers

NASA Headquarters 10

Ames Research Center 14

Dryden Flight Research Center 16

Goddard Space Flight Center 18

Johnson Space Flight Center 20

Kennedy Space Center 22

Langley Research Center 24

Lewis Research Center 26

Marshall Space Flight Center 28

Stennis Space Center 30

and the Jet Propulsion Laboratory 32

Technology Transfer and Commercialization

Overview 36

Applications Projects 38

Network 40

Affiliations 42

Information Resources 44

Commercial Benefits—Spinoffs

Health and Medicine 48

Transportation 54

Public Safety 66

Consumer/Home/Recreation 72

Environment and Resources Management 86

Computer Technology 96

Industrial Productivity/Manufacturing Technology 112

NASA's Commercial Technology Network 120

Aerospace Research and Development

We, NASA, will stick to our vision... We like
stretching the boundaries. We like proving that things
that couldn't be done yesterday will be done tomorrow.
We like making the extraordinary happen.
—*Daniel S. Goldin, NASA Administrator*

For the past 25 years, NASA *Spinoff* has covered successful technology transfer from NASA to private industry. In 1973, then called simply the *Technology Utilization Program Report*, the first issue of the annual publication featured a story on an automated bacteria detection system derived from one used on NASA's Mars-Voyager spacecraft. In this issue, DiaSys Corporation's automated workstation for microscopic analysis also incorporates technology from space research but in a smaller, faster, easier-to-use model. Recycling nonferrous metals from discarded autos featured in 1973 has been updated by the 1997 story on Cryopolymer's application of NASA cryogenic techniques for extracting and recycling rubber from tires.

The original technologies that were eventually transferred to commercial industry would not have existed but for the initial aerospace research and development. Without the Apollo program, cooling systems for the disabled or rechargeable pacemakers would not be the same—or might not exist at all. Materials used in astronaut space suits are now used to protect firefighters on Earth. Technology from the Hubble Space Telescope makes breast biopsies quicker and more accurate, and helped scientists decipher previously unreadable portions of the Dead Sea Scrolls.

Although the Space Station is still a twinkle in NASA's eye, spinoffs from research for long-duration space travel are already appearing. Studies of algae as a food source, recycling agent and oxygen source resulted in a highly nutritious baby formula. Materials research for the space station led to metals used in faucets that prevent scalding from hot water; as illustrated in this issue, these materials also offer improvements in golf clubs.

This section gives a brief description of what NASA Headquarters and each of the ten field centers are doing to contribute to American scientific and technological growth, comparing the years 1973 and 1997 in tribute to the 25th year of spinoff coverage. Just as the Apollo program, Mars-Voyager, and past NASA expertise have led to practical benefits on Earth, it is likely that the current research and development featured in these next pages will lead to tomorrow's revolutionary new commercial products.

NASA Headquarters

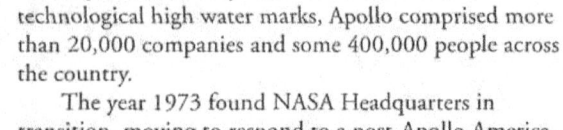

W hen the first artificial satellite of Earth slipped across a backdrop of stars on October 4, 1957, it was heralded in the United States not as a triumph of science and technology, but a bold, startling challenge to America's ideological standing in the world community of nations.

The former Soviet Union's Sputnik 1 satellite sparked a U.S. response, motivating the U.S. Congress to hammer out in early 1958 the National Aeronautics and Space Act. Signed into law on July 29 by then President Dwight Eisenhower, the Act transformed the existing National Advisory Committee for Aeronautics (NACA) into a U.S. civilian space enterprise. That enterprise was named the National Aeronautics and Space Administration (NASA).

On October 1, 1958, just short of a year after Sputnik 1 was cast into space, NASA officially began to blueprint the nation's space program. A fledgling NASA plan of action included human space exploration as well as an agenda of robotic exploration of the Moon and neighboring planets. Hungering for leadership in space exploration and aeronautics, NASA rapidly developed skills and abilities second to none. This striving for excellence led to the creation of NASA field centers spread across the country.

Considered NASA's crowning achievement in the late 1960s was the first human footfall on the surface of the Moon. But perhaps of greater significance was the workforce required to place astronauts on that alien world. Translating Apollo from rhetoric to reality took a NASA managed team of highly skilled government,

industry and university talent. As one of NASA's technological high water marks, Apollo comprised more than 20,000 companies and some 400,000 people across the country.

The year 1973 found NASA Headquarters in transition, moving to respond to a post-Apollo America, a time frame of budget constraints, alternative national priorities, and a seemingly sudden retrenchment in the country setting any new course in space.

The Skylab project was, in effect, a place holder, using remaining Apollo and Saturn V hardware. This experimental space station was rocketed into orbit in 1973, later to be followed by sets of astronauts who lived and worked aboard the facility throughout the year and into early 1974. Skylab became a cost-effective interim substitute for NASA's hoped-for long duration, permanent space laboratory.

For NASA Headquarters, political and White House backing had melded to support development of a reusable transportation system for hauling people and cargo into Earth orbit. Approved the year earlier by then President Richard Nixon, work in 1972 was in full-swing on moving a Space Shuttle program from viewgraph presentations to reality. Bolstered by the prospects of detente between the then Soviet Union and the United States, Nixon had also given NASA the green light to proceed with a joint space mission between the two space superpowers. This effort was successfully carried out two years later as the Apollo-Soyuz Test Project.

The year 1973 saw progress along the aeronautical side of NASA. Dryden Flight Research Center carried out the first piloted flight of the X-24B, a craft built to imitate an unpowered landing of a vehicle returning from space to a runway landing.

Launching pads at Cape Canaveral, Florida were busy in 1973 as NASA launched in April the Pioneer 11 interplanetary probe on a fly-by mission to Jupiter and Saturn. Several months later, in November, an Atlas Centaur hurled into space Mariner 10, a spacecraft that years later encountered Venus, then Mercury in the first demonstration of gravity-assist trajectory change.

Now decades beyond Apollo, Skylab, the early Pioneer and Mariner missions, the NASA of today is far different in organization and purpose. The civilian space enterprise born by Cold War rivalry and one-upmanship

A multi-industry team lead by Lockheed Martin Skunk Works was selected by NASA to build and flight-test a sub-scale X-33 technology demonstrator, shown in this computer-generated concept. Incorporating many proven technologies, this concept could boost the United States back into dominance over the worldwide commercial launch industry.

Astronaut Eugene A. Cernan, Commander of the Apollo 17 mission is photographed by Astronaut Schmitt whose photo is reflected in the gold visor. The climax of NASA's achievement in the late 1960s was the first human footfall on the moon.

is no longer. Relevancy to NASA's ultimate stake-holder—the public—is top priority in a budgetary climate that dictates tough choices among many opportunities. NASA is resolute in its obligation to provide its customers excellent products and services in the most cost-effective and timely manner.

A new NASA has been formed, focused on science and technology programs that enhance and enrich the lives of all Americans. NASA is an investment in the country's future, an agency empowered with a vision to boldly expand frontiers in air and space, inspiring and serving America to benefit the quality of life here on Earth. To attain and sustain this vision, NASA Head-quarters in Washington, D.C. has established an ongoing and iterative Strategic Management Process.

NASA Headquarters provides the organizational structure for the entire space agency. This duty is based on two primary levels of management responsibility. The first is Agency management, which primarily resides at Headquarters. The second is Strategic Enterprise management, which includes managing individual NASA Centers and programs. In essence, NASA Headquarters has become "corporate headquarters." It oversees a set of Strategic Enterprises, and develops NASA's strategy in accordance with what?, why?, and for whom? guidelines that directs the Nation's civil aeronautics and space ventures.

Agency management serves as the principal interface with the stakeholders, including the Administration and Congress. It is the external focal point for accountability, communication and liaison. Agency management provides budget integration, long-term

NASA/stakeholder-focused institutional investment strategy, NASA policy and standards, and Agency functional leadership.

The framework for today's NASA rests on four primary "Strategic Enterprises," each delegated to one or more space agency centers. These are:

Mission to Planet Earth Enterprise:
> Goddard Space Flight Center

Aeronautics and Space Transportation Technology Enterprise:
> Ames Research Center
> Langley Research Center
> Lewis Research Center
> Dryden Flight Research Center

Human Exploration and Development of Space Enterprise:
> Johnson Space Center
> Kennedy Space Center
> Marshall Space Flight Center
> Stennis Space Center

Space Science Enterprise:
> Jet Propulsion Laboratory

These Strategic Enterprises comprise an integrated national aeronautics and space program. Synergism of broad purposes, technology requirements, workforce skills, facilities, and many other dimensions was the basis for amalgamating these activities within NASA in the Space Act of 1958, and the benefits remain strong today.

In further detail, each Enterprise is defined as the following:

Mission to Planet Earth Enterprise

Dedicated to understanding the total Earth system and the effects of natural and human-induced changes on the global environment. The Mission to Planet Earth (MTPE) Enterprise is pioneering the new discipline of Earth system science, with a near-term emphasis on global climate change. Space-based sensors, aircraft, and platforms on Earth offer capabilities presently being used or under development that yield new scientific understanding and practical benefits to the Nation. Today's program is laying the foundation for long-term environment and climate monitoring and prediction. The outcome is the major contribution to the scientific foundation for sustainable development.

Continued

NASA Headquarters *Continued*

Aeronautics and Space Transportation Technology Enterprise

In March of 1997, NASA Administrator Daniel S. Goldin challenged the Aeronautics and Space Transportation Technology Enterprise to set bold objectives for the future. These goals are grouped into three areas, or "Three Pillars:" Global Civil Aviation, Revolutionary New Technology Leaps, and Access to Space. Within the pillars are technology goals, which are framed in terms of a final outcome—NASA-developed technologies being incorporated into industry. Goldin said, "These goals will stretch the boundaries of our knowledge and capabilities. They require taking risks and performing the long-term research and development programs needed to keep the United States as the global leader in aeronautics and space."

The Enterprise will pioneer the identification, development, verification, transfer, application, and commercialization of high-payoff aeronautics technologies. It seeks to promote economic growth and national security through safe, superior and environmentally compatible U.S. civil and military aircraft and through a safe, efficient national aviation system. This Enterprise will work closely in a national alliance with its aeronautics customers, including U.S. industry, the university community, the Department of Defense (DoD), and the Federal Aviation Administration (FAA), to ensure that national investments in aeronautical research and technology are effectively defined and coordinated.

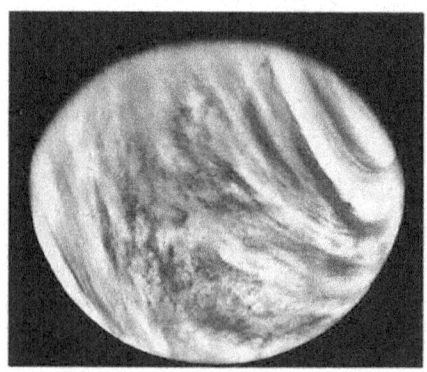

Mariner 10, launched in 1973, photographs Venus the following year. The craft represented the first demonstration of gravity-assist trajectory change.

Lastly, the Reusable Launch Vehicle (RLV) program is developing and demonstrating new technologies for the next generation of space transportation systems. These technologies are moving forward the day when the cost per pound of moving a payload to low Earth orbit can be reduced by an order of magnitude, from $10,000 per pound to $1,000 per pound.

Human Exploration and Development of Space Enterprise

The mission of this Enterprise is to open the space frontier by exploring, using and enabling the development of space and to expand the human experience into the far reaches of space. In exploring space, the Human Exploration and Development of Space (HEDS) Enterprise brings people and machines together to overcome the challenges of distance, time and environment. Robotic science missions survey and characterize other bodies as precursors to eventual human missions. In using space, HEDS emphasizes learning how to live and work there, utilizing the resources and unique environment. The Space Shuttle and the International Space Station pave the way for sustained human presence in space through critical research on human adaptation.

Space Science Enterprise

This Enterprise is multifaceted: science, technology, education and public outreach. The Space Science Enterprise serves the human quest to understand our origin, our existence and our fate. While fulfilling this quest, it seeks to inspire the Nation and the world, to open young minds to broader perspectives on the future and to bring home to every person on Earth the experience of exploring space. Did life arise elsewhere in the universe? Are there worlds around other stars? What is the universe and how did it come into being? This Enterprise addresses these questions and others by establishing a continuum of exploration and science. A virtual presence is to be created in probing new territories within and beyond our solar system.

Workers at Cape Canaveral get one final look at the Mars Pathfinder before it is sealed inside a protective payload fairing for flight. The mission, launched December 2, 1996 arrived at Mars July 4, 1997.

Each of these Enterprises will develop and verify enabling, cutting-edge technologies for future space science, exploration, and commercial missions and will identify and mature high-risk/high-payoff advanced concepts that enable revolutionary new space activities. In addition, they will nurture world-class capabilities that are critical to the development of technologies.

Much cross-cutting of purpose occurs between the Enterprises. The HEDS Enterprise, for instance, provides the Space Science and Aeronautics and Space Transportation Technology Enterprises the means to benefit from human presence in the unique environment of space. Conversely, the Space Science and Aeronautics and Space Transportation Technology Enterprises provide the foundation for the HEDS Enterprise by, among other things, undertaking precursor robotic missions and developing needed knowledge and technology. The Space Science Enterprise enriches the MTPE Enterprise with studies of the Sun, the other planets and the near-Earth environment for their relevance to our understanding of the Earth. The Aeronautics and Space Transportation Technology Enterprise and the HEDS Enterprise are mutually supportive in high-speed aerodynamics, vehicle control systems, and crew accommodation research. These are but a few examples of the mutually beneficial interactions among NASA's Strategic Enterprises.

NASA's Centers are responsible for implementing space agency plans, programs and activities as an integral part of the Strategic Enterprises. Center missions identify the primary concentration of capabilities to support the accomplishment of Strategic Enterprise goals. Meanwhile, Centers of Excellence are focused, agency-wide leadership responsibilities in a specific area of technology or knowledge.

Aligned with NASA's Strategic Enterprises, assigned center missions and the various Centers of Excellence, the space agency is embarking on an exciting and impressive research schedule.

In the area of aeronautics, new strategic goals have been developed with the Federal Aviation Administration (FAA), the Department of Defense (DoD) and industry partners. This combined effort is aimed at a major improvement in the safety of flight, cutting in half the cost of air travel, and equally aggressive reductions in aircraft noise and emissions.

The Origins Program has been established, to look at many facets of the Universe, its creation, the formation of chemical elements and of galaxies, stars, and planets. A major program in Astrobiology is also being pursued, bringing together the best minds in the field of life sciences to research the chain of processes leading to the formation and early evolution of the simplest forms of life in the Universe.

Other highlights of the Origins Program include an accelerated pace for the Mars Surveyor program, assuring a sample return mission from the red planet by the year 2005. A Space Infrared Telescope Facility (SIRTF) is to be launched in 2001, a prelude to building ever larger and more powerful Next Generation Space Telescopes that can observe the Universe to its very beginnings and begin a search to pinpoint Earth-like planets circling other stars.

Excellent progress is being made on the International Space Station. First launchings of key station elements are near at hand. An unparalleled level of experience has been gained in the past few years, working with other nations toward a common purpose. From this first step, new trails can be blazed to other planets and explore the vast expanse of space.

NASA's new Strategic Management Process has been developed as an integrated approach to the planning, implementation, execution, and evaluation of the space agency's activities. This process enables NASA to deliver quality products and services to the agency's customers and stakeholders and to its ultimate beneficiary—the public.

NASA Headquarters has undertaken a major restructuring of the agency's national purpose and has clarified its scientific and technological agenda as the 21st century looms on the horizon. To explore new worlds, advance key technologies, learn to live in space, push the boundaries of flight, understand our changing planet, and inspire America's youth, these are the underpinnings of a revitalized and vibrant NASA.

For the first and probably only time before they are linked together 220 miles up in space, the United States modules for the International Space Station are side-by-side. The Space Station will be a permanent orbiting laboratory in space.

Ames Research Center

Ames Research Center was founded in 1939 as an aircraft research laboratory by the National Advisory Committee for Aeronautics (NACA). While retaining much of its original aeronautics orientation, the center has taken on an increasingly important role in the nation's space program since it became part of NASA when the Agency was founded in 1958.

Situated in Mountain View, California, near San Francisco, Ames was designated the NASA Center of Excellence for Information Technology. In that capacity, Ames's mission is to lead NASA efforts in cutting-edge research in supercomputing, networking, numerical computing software, artificial intelligence, and human factors in order to enable bold advances in both aeronautics and space.

In aeronautics, Ames is the NASA lead in Aviation Operations Systems, championing research efforts in air traffic control and human factors. The center also leads in rotor craft (helicopter) and powered-lift technologies. It has major responsibility in the creation of design and development tools, and for flight simulation, supercomputer consolidation, aeronautical computation and wind tunnel testing.

In space, Ames is NASA's lead center for astrobiology, spearheading research efforts to determine the effects of gravity on living things. Ames plays a major role in efforts to understand the origin, evolution and distribution of stars, planets and life in the universe; in ecosystem and atmospheric science in support of Mission to Planet Earth; and is leading NASA efforts in developing thermal protection systems for future access to space and for planetary atmospheric entry vehicles.

Ames is home to an impressive array of research facilities. This includes three national wind tunnels, including the world's largest; several advanced flight simulators; a variety of supercomputers, including some of the world's fastest; a suite of centrifuges that serve as a national resource; and several unique aircraft—both fixed-wing and rotor craft—used for aeronautical flight research and as flying scientific laboratories.

A look back to 1973 at Ames provides a snapshot of work in progress. In 1973, a joint NASA-Soviet Union study undertook to analyze ice flow, meteorological data and wildlife migration patterns in the Bering Sea. For this study Ames provided a Convair 990, a converted plane that was transformed into a research tool.

Far out in space, the Pioneer 10 spacecraft was hurling toward Jupiter, outward bound after liftoff on March 2, 1972. Operated from Ames Research Center, Pioneer 10 reached Jupiter in December 1973, then went on to become the most distant and longest-lived interplanetary explorer. Pioneer 11 was rocketed into space in April 1973, also bound for Jupiter, later to make history's first trip to Saturn, arriving in 1979.

In March 1997, the hardy Pioneer 10 reached its 25th anniversary in space and a point 6.2 billion miles from Earth. The Pioneer carries a message for any intelligent life forms that it might encounter on its sojourn across the galaxy. Affixed to the spacecraft is a plaque, depicting a man and a woman, a map of Earth's solar system, and other symbols which may help intelligent beings interpret the message, learn about the spacecraft's creators, and where they lived. The Pioneer 10 mission ended in March 1997 when power became too weak to transmit data back to Earth.

Early in the 1970s, the Illiac IV greatly strengthened the center's work in theoretical fluid mechanics, as well as concentrating superior computer strength at the center. Obtaining and locating Illiac IV—then the largest and most sophisticated computer in the world—was a coup for Ames. Thanks to an agreement with the Advanced Research Projects Agency (ARPA), part of the Defense Department, Ames was able to house and manage the computer complex, which was in full operation by early 1973.

America's aeronautics history is an undisputed success story in global competitiveness. To support the vital role of aircraft manufacturing, NASA created the Numerical Aerodynamic Simulation (NAS) facility in 1984. Ames is the locale for this world-class supercomputing capability, accessible to the nation's aeronautical researchers in government, industry and academia. Ames is resolute in developing faster, more

Launched on March 2, 1972 and operated from Ames Research Center, Pioneer 10 reached Jupiter in December 1973. In March 1997, the hardy Pioneer 10 reached its 25th anniversary in space and a point 6.2 billion miles from Earth.

powerful computing systems and to broaden American know-how as a national vision with no limits. The foundation for the work done at NAS is Computational Fluid Dynamics (CFD). This involves using computers to study the fluid flow around an object, such as a wing or rotor blade. NAS supported the designers of most vehicles made in America today, from planes and rockets to cars. Incorporating the Ames CFD program as a tool for researchers has been noted as the start of a genuine revolution, the full consequences of which are still to be felt.

Ames space science researchers quite literally "took the plunge" on December 7, 1995. Galileo's atmospheric entry probe successfully measured Jupiter's atmosphere directly. This probe was developed and managed at Ames. The center's Galileo Probe Project conducted scientific and engineering studies enabling this most difficult atmospheric entry.

The Space Directorate at Ames is coordinating NASA research in the broad discipline called astrobiology, a relatively new term for studies of life in the universe. By way of Ames's Center for Mars Exploration, future Mars missions are being assessed, with emphasis on the search for life.

Major flight projects underway at Ames include biological research facilities for the International Space Station; Stratospheric Observatory for Infrared Astronomy (SOFIA), a flying airborne astronomical laboratory with a 100-inch infrared telescope; Lunar Prospector, the first NASA mission to study the Moon since Apollo; and thermal protection system technology for today's Space Shuttle and tomorrow's single-stage-to-orbit, fully-reusable launchers.

Being centered in an area with one-fifth of the 100 fastest-growing international companies has offered Ames a distinct benefit in maintaining its outlook on future technologies. Achieving immense computing capacity is a critical national priority, and Ames has between 30 and 40 percent of NASA's entire supercomputing capability. This important national objective is perhaps best personified in the NASA High Performance Computing and Communications (HPCC) Program.

One measure of its computer competence is the center's lead role in the research and development efforts on the Next Generation Internet (NGI) initiative. By 2002, this initiative could result in information flowing a million times faster than today's modern home computer modems and 1,000 times faster than a current standard T1 business computer line.

As lead institution for NASA's portion of a three-year, $300 million federal project to develop the NGI, Ames joins other federal agencies in the venture,

including the National Science Foundation, the Defense Advanced Research Projects Agency, the Department of Energy, the National Institutes of Health, and the National Institute of Standards and Technology (NIST).

Ultimately, the NGI should guarantee levels of service that would dissolve slowdowns and network stagnation. These slowdowns can make waiting for Internet images, movies and other services a tedious, time-consuming process. NGI would tie core sites with high speed lines, leading to GigaPOP interconnectivity across the country. GigaPOP is a regional group of core organizations that will connect their separate computer network systems by high speed communications lines. A POP is a point of presence. Giga stands for a billion computer bits.

An early step for the NGI is linking up some 100 universities, research labs and other institutions at a hundred times the speed of today. GigaPOP interconnects would continue to be enhanced to handle computer data at ever speedier rates, bringing about an Internet upgrade that can function at much higher speeds than today.

Ames Research Center's mission has grown as national needs for research and technology contributions have broadened. From civil and commercial aviation research, empowering users with the ability to move volumes of data via computer to exploring the very origins of life itself—these serve as tenets of a NASA research center poised for the 21st century.

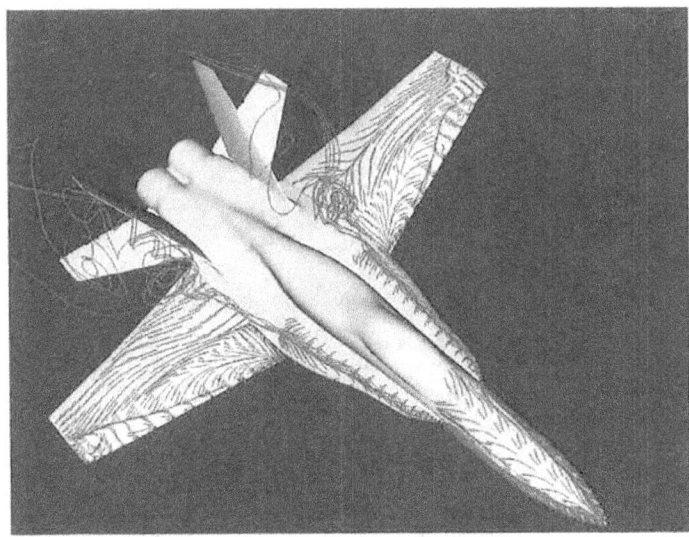

Computational fluid dynamics are used to determine the effect of sideslip on the flow around an aircraft. Supercomputer design has the potential to be less costly and gives data not available from wind tunnels, such as greater detail on flight conditions.

Dryden Flight Research Center

Some of the nation's greatest advances in aviation have been staged at Hugh L. Dryden Flight Research Center. It is the home of aircraft that push the envelope of aerodynamic theory, along with research pilot skill, nerve, and courage.

Dryden is located at Edwards, California, in the Mojave Desert, some 80 miles north of Los Angeles. Enjoying nearly perfect weather for flight research, this NASA facility sits at the southern end of a 260-mile high-altitude high-speed supersonic corridor (there are three high-speed corridors altogether). Situated adjacent to Rogers Dry Lake, a 44-square mile natural surface for landing, Dryden is in an isolated area free of population density.

Dryden has data catalogued from years of flying experimental aircraft, such as the Bell X-1—the first aircraft to break the sound barrier on October 14, 1947. The X-1 program provided the National Advisory Committee for Aeronautics (NACA), NASA's predecessor, an ability to master the effects of transonic speeds

and the stability and control of aircraft flying in that regime.

Since the late 1940s, Dryden has acquired a unique and highly specialized capability for conducting flight research programs. Its research organization, consisting of pilots, engineers, technicians and mechanics is unmatched anywhere in the world.

The open skies, land and resources at Dryden proved their usefulness to the space effort in a large way. The Mach-6 X-15 program researched and developed various technologies that were implemented in the U.S. Mercury, Gemini and Apollo spacecraft. The X-15 provided the pioneering work needed to design a craft to go into space, then return to a horizontal landing on Earth. Along with the X-15, lifting body research done at Dryden in the 1960s helped pave the way for today's Space Shuttle.

In September 1996, Dryden celebrated its 50th anniversary of contributing to the nation's aerospace capabilities. This expertise was clearly in evidence in 1973 as the first flight of the X-24B took place. The X-24B was successor to the wingless lifting bodies that were flown in a joint USAF-NASA research program to demonstrate a pilot's ability to maneuver and safely land a vehicle with a shape designed for reentry from space flight.

Like its wingless lifting body predecessors—the M-2, HL-10 and the X-24A that had flown at Dryden—the X-24B made use of a B-52 for air launch above the dry lake bed. Released at 40,000 feet, the X-24B glided to a desert touchdown a scant four minutes later. Flown from August 1973 to November 1975, the X-24B proved invaluable in shaping the Space Shuttle program.

An F-8 Crusader first flown at Dryden in 1972 inaugurated in 1973 the first tests of Digital Fly-By-Wire (DFBW), a concept that utilizes an electronic flight control system coupled with a digital computer. The research aircraft tested DFBW as a replacement for conventional mechanical flight controls. Well over 200 flights of the F-8 Crusader were carried out in a DFBW program that lasted 13 years—considered one of the most significant and successful aeronautical programs in NASA history. Fly-by-wire flight control technology, made possible in large measure by the F-8 Crusader tests, was later applied in creating the Space Shuttle flight control system.

The center's primary study tools are research aircraft. But ground based facilities play a significant role in Dryden research. These key facilities include a high temperature and loads calibration laboratory to test complete aircraft and structural components under the

The X-24B is surrounded by support personnel on Rogers Dry Lakebed after a research mission as two F-104s and a T-38 fly over head. First flown at NASA Dryden in 1973, the X-24B demonstrated the ability for a pilot to maneuver and safely land an unpowered craft on a conventional runway, as the Space Shuttle does today.

combined effects of loads and heat, a highly developed aircraft flight instrumentation capability and a flight systems laboratory with a diversified capability for avionics system development.

Other assets of Dryden include a flow visualization facility that permits examinations of how air courses around test models or small components, a data analysis facility for processing of flight research data and a remotely piloted research vehicles facility.

Dryden participated in the five free-flight Approach and Landing Tests of the Space Shuttle Enterprise staged in 1977 and continues to support Shuttle orbiter landings if diverted by bad weather from the Kennedy Space Center in Florida. If a shuttle does land in California, the vehicle is then ferried back to the launch site atop a 747 aircraft.

Clearly in evidence at Dryden is the match of the center's past prestige with the necessary technical competence to tackle the aeronautical challenges of the day. In this regard is the specially instrumented F/A-18, used to investigate high angle of attack, or high alpha, flight. Today's high performance jet aircraft can fly in the high alpha flight regime, but not necessarily efficiently. The center's research created a data base for aircraft designers to accurately predict high alpha airflow aircraft control and performance. High alpha technology may result in better control and maneuverability and enhanced safety in future high performance aircraft.

Another high alpha program at Dryden featured the thrust-vectored X-31. An international test organization managed by the Advanced Research Projects Agency (ARPA) conducted flight tests to obtain data for next generation high performance aircraft. In addition to NASA and ARPA, program participants

included the U.S. Navy, U.S. Air Force, Rockwell Aerospace, Deutsche Aerospace, and the Federal Republic of Germany.

In 1993 and 1995, a Propulsion Controlled Aircraft (PCA) system underwent successful tests at Dryden. A PCA system provides a pilot with a computerized system to land an aircraft with only engine controls in the event of a catastrophic hydraulic system failure. Once considered an impossible feat by many engineers, automatic PCA landings of two McDonnell Douglas aircraft, an F-15 fighter and an MD-11 airliner, were accomplished.

The Environmental Research Aircraft and Sensor Technology program at Dryden is attempting to develop remotely controlled aircraft capable of sustained, slow flight at high altitudes to gather currently unavailable information about our atmosphere. For example, Pathfinder is a solar-powered, ultra light research aircraft developed by AeroVironment Corporation. The vehicle will test very high-altitude and extremely long-duration flight for periods of up to several weeks or months. Key areas of development include solar cell, battery and electric motor technology; flight operations techniques and procedures; structures; flight environment simulation; and science mission demonstration.

For the past half century, Dryden Flight Research Center has been at the forefront of flight research; a place where people can be engaged in examining and resolving the great aeronautical and astronautical challenges of our time. By pushing the experimental envelope, be it speed, altitude, control, or other boundaries, Dryden research has strengthened the U.S. position as a world-class leader in aeronautics.

NASA F-8 Digital Fly-By-Wire research aircraft is on one of its pioneering flights in the 1970s. Dryden-led research developed electronic fly-by-wire flight control systems used in many aircraft today, including the Space Shuttle orbiters.

Goddard Space Flight Center

W hen you drive through Goddard Space Flight Center in Greenbelt, Maryland, you see history at every corner. A case in point is the intersection of Minitrack and Tiros roadways. These names were part of Goddard's early space history in tracking and obtaining early images of Earth's weather.

Dotting a sprawling 1,200-acre complex of buildings are control centers, geophysical and astronomical observatories, as well as antenna and laser test ranges. Goddard has been at the forefront of U.S. space progress since it was established in January 1959. Ten miles northeast of Washington, D.C., Goddard has one of the world's leading groups of scientists, engineers and administrative managers. It has the largest scientific staff of all the NASA centers.

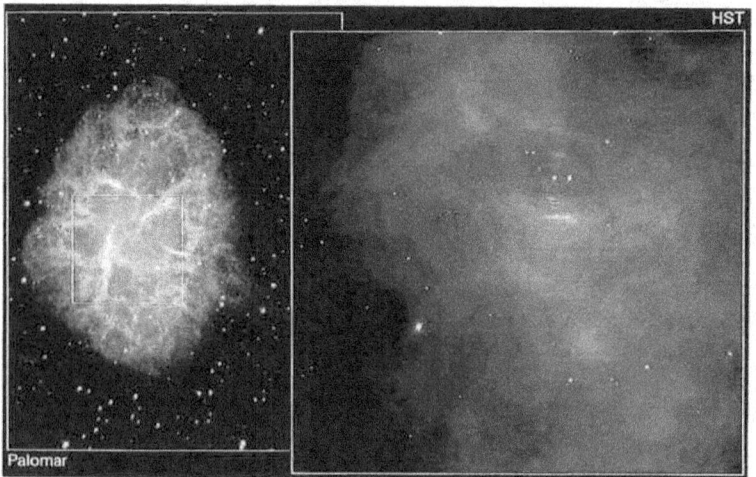

A new sequence of Hubble Space Telescope images of the remnant of a tremendous stellar explosion is giving astronomers a remarkable look at the dynamic relationship between the tiny Crab Pulsar and the vast nebula it powers. Left is the entire Crab Nebula; right is an image of the inner parts of the Crab.

The center is named after American rocket pioneer, Robert Goddard, who designed and flew the world's first liquid-fueled rocket. NASA established Goddard as its first major scientific laboratory devoted entirely to the exploration of space.

Because of its versatility, Goddard scientists can develop and support a mission, while Goddard engineers and technicians design, build and integrate spacecraft. Goddard is also involved in implementing suborbital programs using aircraft, balloons, and sounding rockets. This function is located at the Wallops Flight Facility on Wallops Island, Virginia.

Much of the center's theoretical research is carried out at the Goddard Institute for Space Studies in New York City. Operated in close association with area

universities, the institute provides support in geophysics, astronomy and meteorology.

Goddard has played a key role in space progress almost from the day it opened. From the early Vanguard and Tiros weather watching satellites to the Compton Gamma Ray Observatory, the Cosmic Background Explorer, the Upper Atmosphere Research Satellite, the Extreme Ultraviolet Explorer and controlling the Hubble Space Telescope, Goddard is a site of active and vibrant scientific and technological research. Goddard is a preeminent world leader in Earth and space science. Scientific research in Earth science, physics and astronomy is its designated Center of Excellence area.

Rolling back the clock to 1973, Goddard's seagoing tracking station, the USNS Vanguard, was sailing in late-March to its new temporary home in Mar Del Plata, Argentina. As part of NASA's Spaceflight Tracking and Data Network, the ship was moved from its home port near Cape Canaveral, Florida to support the forthcoming Skylab space station launch.

Goddard's Earth Resources Technology Satellite (ERTS) team was receiving kudos for its involvement in the program to use a spacecraft that studied crops and scanned for resources. Launched in July 1972 as the world's first Earth remote sensing satellite dedicated to resource mapping the globe, ERTS-1, was later renamed as part of the Landsat series. By November 1973, Goddard's NASA Data Processing Facility had processed and shipped its one millionth film product from the ERTS.

That same year, the Computer Software and Management Information Center (COSMIC®) in Atlanta, Georgia had selected several Goddard computer programs for distribution. These computer programs ranged from a unique system for automatically wiring tiny electrical circuits to a mechanistic numerical program "debugging" tool for use with conventional computer languages.

Scientists from around the world attended a May 1973 workshop on Gamma-Ray astrophysics. Symposium topics covered, in part, the changing picture of the universe made possible by experiments aboard Apollos 15,16, and 17. At the same gathering, preliminary results were offered from the second Small Astronomy Satellite, launched in November 1972.

Also in 1973, center scientists were ecstatic about the liftoff of the Goddard-built Radio Astronomy Explorer-B. Renamed Explorer 49, the spacecraft sent back first scientific data from a near perfect "anchor'" orbit around the Moon. Meanwhile, throughout the year, meetings were underway between Marshall Space Flight Center and Goddard on using the Space Shuttle in early 1980 to launch the Large Space Telescope.

Lastly, Goddard space scientists were pondering how best to capture a rare visitor from space. Comet Kohoutek was to make its closest approach at year's end, with ground and spaceborne equipment at the ready to study the celestial passerby.

Moving the clock forward to today, Goddard Space Flight Center is committed to excellence in scientific investigation, in development and operation of space systems, and in advancement of essential technologies.

This commitment is evidenced by the highly successful, Goddard-managed, servicing missions of the Hubble Space Telescope (HST) in December 1993 and in February 1997. Goddard is also the home of the Space Telescope Operations Control Center (STOCC). The STOCC is the nerve center for HST where all commands for the telescope originate.

Goddard is also responsible for procurement, development and verification testing of the Geostationary Operational Environmental Satellite (GOES). Once checked-out, each GOES is turned over to the National Oceanic and Atmospheric Administration (NOAA).

A host of other satellites are now waiting in the wings, being prepared for launch. Submillimeter-Wave Astronomy Satellite (SWAS) is one of three missions in NASA's Small Explorer program. The Goddard-managed SWAS is built to investigate one of the great mysteries of astronomy: how stars are born.

Sea-viewing Wide Field-of-view Sensor (SeaWiFS) is considered to become a powerful tool for understanding the biological and physical processes in the world's oceans by measuring the color of the ocean. This instrument will be the only scientific payload aboard the SeaStar commercial spacecraft. SeaWiFS will provide large amounts of ocean information to the Earth science community.

The Tropical Rainfall Measuring Mission (TRMM) is a joint scientific satellite project between NASA and the Japanese Space Agency. The TRMM office, based at Goddard Space Flight Center, sponsors several rainfall research programs in the United States and abroad.

Arguably, the most important venture that can be undertaken in space is NASA's Mission to Planet Earth (MTPE). Goddard is lead center for the MTPE program. The Earth Observing System (EOS) is the centerpiece of MTPE and is managed by Goddard. EOS will feature a series of polar-orbiting and low inclination satellites for global data-collection.

The end product of MTPE will be the ability to develop and implement environmental policies based on a better understanding of how our environment works. To develop that understanding, MTPE will rely on the EOS Data and Information System (EOSDIS). The EOSDIS

has been designed to achieve, manage and distribute MTPE data worldwide.

Just as the first weather and resource monitoring, space astronomy and communications satellites fundamentally changed our way of thinking about those fields, so the element of Mission to Planet Earth will expand our perspective of the global environment and climate. By harnessing data gleaned by satellites, better stewardship of planet Earth is possible.

In no small measure, Goddard Space Flight Center's role in these efforts underscores what Robert Goddard said many years ago: "There can be no thought of finishing, for 'aiming at the stars,' both literally and figuratively, is a problem to occupy generations, so that no matter how much progress one makes, there is always the thrill of just beginning."

® COSMIC is a registered trademark of the National Aeronautics and Space Administration.

The Goddard-built Radio Astronomy Explorer-B (renamed Explorer 49) was launched in June 1973 from Cape Kennedy by a Delta vehicle. The Explorer mission's goal was to detect and measure the intensity of low frequency radio signals from solar, galactic and extragalactic sources.

Lyndon B. Johnson Space Center

Putting a face on space exploration first meant short excursions above Earth by an astronaut sitting alone in a small, confining "capsule" who ended a brief journey under a billowing parachute and plopping down into cold, ocean waters.

Astronaut flight of today features a crew that wings their way into Earth orbit in a huge space shuttle, docking with a space station, deploying scientific spacecraft or servicing a large observatory, then gliding to a smooth, runway touchdown at mission's end.

Decades of pacesetting work has created a road into space for human travelers that stretches into Earth orbit and has carried humans to the faraway shores of the Moon.

NASA's designated Center of Excellence in human operations in space is Lyndon B. Johnson Space Center in Houston, Texas. Through the Mercury, Gemini, Apollo, Skylab, Space Shuttle, and the planned International Space Station, the Johnson Space Center has been the hub for extending and expanding human operations in space.

Established in September 1961, first as the Manned Spacecraft Center (MSC), Johnson Space Center is NASA's primary center for design, development and testing of spacecraft and associated systems for human space flight. This center selects and trains a cadre of astronauts for space missions, and is extensively involved in medical, engineering and scientific experiments carried aboard space flights.

Johnson's Mission Control Center is where all human space flights are monitored, located at the center of the complex. A Space Station Control Center was completed in November 1991, and is now ready to support the International Space Station project.

Johnson is the very nerve center of America's piloted space endeavors. No doubt, many a television viewer has watched the cool faces of Mission Control Center workers experiencing a range of emotions, be it from "Houston, Tranquility Base here. The Eagle has landed," to "Houston, we've got a problem..."

Since 1965, the Mission Control Center has been vital to the success of every piloted space flight since the Gemini 4 mission. These teams of experienced engineers and technicians monitor systems and activities aboard spacecraft 24-hours a day during missions, using sophisticated communication, computer, data reduction and data display equipment. Watching each movement the crew and spacecraft make, Mission Control personnel double-check every number to assure that missions are proceeding as planned, and provide the expertise needed to deal with the unanticipated.

A complex of some 100 buildings dot the Johnson site of 1,620 acres near Clear Lake. The center has

The Space Shuttle Discovery on its Mobile Launch Platform makes its slow 3.4-mile trek from the Vehicle Assembly Building to the launch pad in preparation for the STS-82 mission. The mission crew performed the second servicing of the orbiting Hubble Space Telescope in early 1997.

management responsibility for the Space Shuttle program, as well as a major accountability for the development of the International Space Station (ISS). The center is responsible for the interfaces between the ISS and the Space Shuttle and flight operations of both, and maintains excellence in the fields of space systems, engineering, life sciences, and lunar and planetary geosciences.

Additional Johnson facilities are located at nearby Ellington Field, with the center also responsible for direction of operations at the White Sands Test Facility. These facilities are situated on the western edge of the U.S. Army White Sands Missile Range at Las Cruces, New Mexico, supporting the spacecraft propulsion check-out, power system evaluations and materials testing.

Johnson Space Center engineers and scientists in 1973 were busy debriefing the last set of Apollo astronauts sent to the Moon. The Apollo 17 crew had completed a 12-day mission December 19, 1972 ending the lunar landing program initiated by then President John F. Kennedy in 1961. After six lunar landings in the Apollo program, twelve people had walked on the Moon's surface, explored the "magnificent desolation" as moonwalker Buzz Aldrin called it, and returned 842-pounds of lunar rock and soil samples. During 1973, this precious cache of resource material was under intensive analysis at Johnson's Lunar Receiving Laboratory and being distributed to anxious scientists around the world.

From May 25 through February 8, 1974, three missions of astronaut teams to Skylab were completed. Each of those flights consisted of a three-person crew. Over this period of time, engineering groups on the ground were mobilized to problem solve. Through Johnson's Mission Control, obstacles were overcome as Skylab crew members fixed the station's woes, thus enabling the astronauts to chalk up months of living and working aboard the orbiting complex.

Early in 1973 the name change from the then Manned Spacecraft Center (MSC) to today's Lyndon B. Johnson Space Center was executed. This dedication was fitting the vision of former President Lyndon B. Johnson who helped draft, introduce, and enact the legislation that created NASA.

Lastly, 1973 was also a time when planning was underway for the first joint undertaking with the former Soviet Union. The Apollo-Soyuz Test Project to be flown two years later would mark a new era of cooperation between spacefaring nations, sowing the seeds for U.S.-Russian collaboration in evidence today. While busily at work on this mission, Johnson Space Center planners were orchestrating America's next generation of space transportation, the Space Shuttle, a just-approved project by then President Richard Nixon.

No longer would launching cargo into space be a one-way trip. Led by Johnson Space Center engineers, the Space Shuttle Transportation System would consist of a fleet of piloted space planes called "orbiters," mounted to a giant external fuel tank and two solid rocket boosters. The result is a new class of space travel, a revolutionary way to gain access to Earth orbit.

Among the specialized training facilities at Johnson are Shuttle simulators; Space Shuttle Orbiter Trainer, the Manipulator Development Facility, Precision Air Bearing Facility and Space Station mockups; and the Weightless Environment Training Facility.

Life sciences, planetary and Earth sciences, robotics, artificial intelligence and lunar samples are a few of the center's research areas dedicated to space and life sciences.

Engineering facilities at Johnson Space Center include vacuum chambers, an anechoic chamber, antenna range, avionics testing and various structural and environmental test areas. These and other facilities, hardware and test equipment are brought to bear to support human space missions using the Space Shuttle today, and in the future as the era of the International Space Station matures.

Introduced into service in 1981 as the world's first reusable space vehicle, the Space Shuttle system has carried hundreds of people into space and delivered several millions of tons of cargo into orbit. The Space Shuttle continues today as the nation's most capable form of space transportation. It is supporting flights of astronauts to the Russian Mir space station, a prelude to assembling the International Space Station to be finished in the opening years of the next century.

In the nation's quest to master space, Johnson Space Center is positioned to seek new answers to new questions in the arena of living and working in the space environment. That work will entail building large space structures, supporting an eventual human return to the Moon and dispatching expeditions to the distant dunes of Mars.

Through the guidance of Johnson Space Center's Mission Control, astronauts chalked up months of living and working aboard Skylab beginning in 1973.

Kennedy Space Center

Riding a plume of flame, a Space Shuttle rumbles through Florida skies and heads for Earth orbit. This scene has become a common occurrence at the John F. Kennedy Space Center, America's premier spaceport.

As NASA's designated Center of Excellence in launch and cargo processing systems, Kennedy Space Center has a mission area of space launch...and for good reason. This NASA center has long been a historic departure point for both expendable rockets and human space travelers.

The Kennedy Space Center is situated on Florida's central Atlantic coast, carved out of savanna and marsh in the early 1960s. Determined as ideal for launchings and landings, Kennedy's "space coast" real estate evolved from a sandy strip 34 miles long and five to 10 miles wide on Florida's east cost, midway between Jacksonville and Miami.

This center occupies 140,000 acres of land and water on Merritt Island. Thanks to Kennedy's large area and surrounding water, ample safety is provided to the surrounding communities during launches, landings and other hazardous operations. Only a small portion of Kennedy is used for space operations; the balance is managed by the U.S. Department of the Interior as a wildlife refuge and national seashore.

Early piloted missions in the Mercury and Gemini series took off from the Eastern Test Range, also known as Cape Canaveral, adjacent to where the Kennedy Space Center is now located. Prompted by an American commitment in the 1960s to conquer the new ocean of space, NASA began acquiring land across the Banana River from Cape Canaveral in 1962.

By January 1963, work was in full swing on an all-NASA "Moonport" on Merritt Island. These were heady times for the newly-built Kennedy Space Center. The drama of the so-called "Space Race," Cold War competition between the U.S. and the former Soviet Union, manifested itself with the Moon becoming an undeclared finish line.

Everything about developing the Kennedy Space Center was done in large scale—similar in stature to the daunting challenge at the time: placing Apollo astronauts on the Moon and returning them safely to Earth by the close of the 1960s.

Three-and-a-half miles distant from Launch Complex 39, engineered to handle the awesome power of Saturn liftoffs, a 525-foot high structure was built to assemble the mammoth boosters. This structure, now tagged the Vehicle Assembly Building, towered above the Florida landscape. It was capable of erecting four of the monstrous Saturn V boosters simultaneously. To move a fully stacked Saturn V to its launch pad, a giant diesel-powered crawler transporter would lumber across a specially-built roadway. A Launch Control Center served as the brains of the Moonport, directing mission support, fuel loading, and launch of the powerful Saturn launch vehicles.

Construction activities lead to an operational Launch Complex 39 by 1967. Twelve piloted and unpiloted Saturn V/Apollo missions rocketed away from Kennedy Space Center between 1967 and 1972.

In May 1973, the Skylab space station was tossed into orbit by a Saturn V from the Kennedy Space Center, later to be followed by a trio of separate launchings of three-member crews using the smaller Saturn IB rockets. The Saturn/Apollo era ended in 1975 with the liftoff of an Apollo crew on a Saturn IB toward a joint docking with cosmonauts aboard a Soyuz spacecraft launched by the former Soviet Union.

As the Apollo program concluded, Kennedy began its transformation for sustained Space Shuttle operations. The 1970s saw modifications to existing facilities

Looming over the landscape like a sleeping giant, the Vehicle Assembly Building towers 525 feet above the terrain at Kennedy Space Center. Built for the Apollo program in the 1960s, it is now the final assembly point for the Space Shuttle.

to inaugurate the Shuttle era. For example, in 1979, a three-mile long Shuttle Landing Facility and an Orbiter Processing Facility were built. New checkout and launch procedures were developed. Entirely new sets of computers, called the Launch Processing System, were installed in the Launch Control Center. The huge Vehicle Assembly Building was converted to handle Shuttle components, while crawler transporters were changed to suit Space Shuttle configurations.

At Launch Complex 39, the spot where rockets once lifted off bound for the Moon, engineers altered facilities to handle the business end of launching Space Shuttles. No longer would a vehicle be prepared for a single flight. The same vehicles would return again and again to the Kennedy Space Center, to be processed and made ready for flight once more.

The Kennedy Space Center conversion from Apollo to the Space Shuttle effort reached a major milestone with the start of an Orbital Flight Test Program, marked by the maiden flight of the first Space Shuttle mission on April 12, 1981.

Today, and over 80 Shuttle missions later, the Kennedy Space Center continues as the locale for Shuttle integration and rollout, payload processing, prelaunch checkout, launch pad operations, launch, recovery, and ground turnaround operations.

Kennedy Space Center's responsibility also extends to the facilities and ground operations at Vandenberg Air Force Base in California and designated contingency landing sites. Furthermore, Kennedy is NASA's prime center for payload testing and checkout and provides oversight of space agency missions launched on expendable rockets from Cape Canaveral Air Station, Florida. These cargoes could be a scientific satellite headed for Earth orbit, or bound for interplanetary targets such as the ringed Saturn.

Over the brisk pace of NASA's early beginnings, Kennedy has advanced launch procedures, facilities and equipment and cultivated special skills to put the United States into space. Now that the space frontier has been crossed by humans, from Earth orbit to the Moon, there is no turning back. Human curiosity is demanding, as is the response in science and technological development.

Technology inexorably prods the U.S. space program forward, in search of new horizons. Yet another foothold on the future is the International Space Station (ISS). This permanently crewed orbiting facility can enable the U.S. and its partners in Canada, Europe, Japan and Russia to use the space environment to the fullest for the benefit of humankind.

The ISS demands a magnitude of effort and commitment comparable to Apollo/Saturn and Space

Shuttle. Once again, new initiatives are required to process and store large, complex space structures and prepare them for flight. The Space Shuttle is integral to the success of the ISS as it will ferry into space the people and materials necessary to piece together and eventually operate the immense orbiting facility. Logistical integration of station elements with on-going Space Shuttle operations at the Kennedy Space Center pose formidable technical and managerial challenges.

Moreover, as the Shuttle fleet nears the end of its design life, shortly after the turn the of the century, a new generation of lower-cost space transportation for passenger and cargo is to come on line. Drawing from its historic roots, the Kennedy Space Center will bring to this new venture decades of hard won experience to streamline space vehicle checkout, turnaround and safe launch operations.

The outlook for space is indeed exciting and challenging.

As the 21st Century approaches, such grand achievements as orbital industries, Moon bases and human treks to Mars and beyond may be realized.

Whatever paths are taken in the tomorrows yet to come, they are to be founded in large part on the unique experience and resources of the Kennedy Space Center—America's Spaceport to the future.

An electrical connector, developed by Kennedy Space Center engineers, is dwarfed in size by a dime. The connector was developed for use by the Rancho Los Amigos Hospital in Downey, California, for the treatment of paralyzed muscles in 1973.

Langley Research Center

angley Research Center, established in 1917 as the first national civil aeronautical laboratory, has been shaping U.S. aeronautical and space prowess for some 80 years.

Situated in Hampton, Virginia, Langley served as the first research laboratory for NASA's predecessor, the National Advisory Committee for Aeronautics (NACA). In this unique role, the center bore basic responsibility for nourishing U.S. aviation from infancy to world leadership. Dozens of wind tunnels and other distinctive facilities are at the disposal of researchers to aid in the investigation of the full flight range—from general aviation and transport aircraft to hypersonic vehicles and reusable rocketry.

Langley is NASA's Center of Excellence in airframe systems and leads in airborne systems, structures and materials, aerodynamics, mission and systems analysis,

Conducted jointly by Langley and Dryden Flight Research Center, the Hyper-X program will demonstrate technology that could ultimately be applied to vehicles from hypersonic aircraft to reusable space launchers. When the Hyper-X flies, it will be the first time a non-rocket engine has powered a vehicle in flight at hypersonic speeds.

and crew station design and integration. The center leads the Agency's Advanced Subsonic Technology (AST) program and the NASA-industry High Speed Research program (HSR). Langley also leads efforts in hypersonic propulsion, and is one of four centers that is integral to NASA's Aeronautics and Space Transportation Technology Enterprise. A goal is to develop technologies to enable aircraft to fly faster, farther, safer and to be more maneuverable, quieter, less expensive to manufacture and more energy efficient.

In 1973, one major effort at Langley was orchestrating NASA's research into use of LIDAR (laser-radar) to monitor atmospheric pollution. The overall purpose was to measure the aerosols at various levels of the lower atmosphere. By studying the "residence" times of the aerosols at particular sites, the movement and redistribution of particles over the long term could be assessed.

Upon arrival of the 1973-1974 energy crisis that swept across America, Langley also sought to identify any and all ways to use airplane fuel more efficiently. This endeavor was manifested in the Aircraft Energy Efficiency (ACEE) program. An inventory of then-available and future technologies that could be used by aircraft manufacturers was a thrust of ACEE. Langley's specific research for ACEE was in the areas of materials, structures and aerodynamics.

Casting an eye toward the passenger jet of the future, Langley is now working on next-generation supersonic aircraft capable of moving 300 people at more than 1,500 miles per hour—more than twice the speed of sound. The High-Speed Civil Transport (HSCT) would cross the Pacific or Atlantic in less than half the time of modern subsonic jets. Better yet, the travel would be affordable—at a ticket price less than 20 percent above comparable, slower flights!

HSR is supported by a team of major U.S. aerospace companies in a multi-year research program that started in 1990. The international stakes are high, and for good reason. It is estimated that the market for more than 500 HSCTs between the years 2000 and 2015 translates to more than $200 billion in sales. There is the potential for 140,000 new jobs in the United States to bring about HSCT development.

Langley's High-Speed Research initiatives also consists of the "SuperVIEW" cockpit. SuperVIEW stands for the Supersonic Video Integrated Electronic Window. Forward cockpit windows in future supersonic passenger aircraft may be replaced by large displays with video and infrared images, enhanced by computer-generated graphics. SuperVIEW displays would guide pilots to the airport, warn them of other aircraft near the flight path, and provide cues for airport approaches and landings.

Another initiative is the Hyper-X program, an Agency-wide effort to address one of the greatest aeronautical research challenges—hypersonic flight. Conducted jointly by Langley and Dryden Flight Research Center, program managers hope to demonstrate technology that could ultimately be applied to vehicles from hypersonic aircraft to reusable space launchers. MicroCraft, Inc., of Tullahoma, Tennessee, was selected to fabricate a series of small, unpiloted experimental vehicles that will fly up to ten times the speed of sound. When the Hyper-X flies, it will be the first time a non-rocket engine has powered a vehicle in flight at hypersonic speeds—speeds above Mach 5, equivalent to about one mile per second or approximately 3,600 miles per hour at sea level. A booster rocket will carry each experimental vehicle to its flight-test speed and altitude, where it will be launched to fly under its own power.

Langley also plays a critical role in short-circuiting the decline of general aviation. A revitalization of general aviation—all flying except the military services and commercial airlines—is a must, as is recovering a leadership position in this arena. To this end, an Advanced General Aviation Transport Experiments (AGATE) consortium is directed by the Langley head of a general aviation office. AGATE is a partnership between government, industry and academia to stem the gradual deterioration of general aviation in the United States.

An assignment Langley is undertaking clearly draws from decades of aeronautical work. New and innovative technologies are being explored to expand piloted and unpiloted space flight, but at the same time, lower the cost of routine access to orbit. Similar to work done in support of the now-flying Space Shuttle, Langley researchers are assisting industry to develop and introduce the next generation of space vehicles. One top priority is a fully reusable spacecraft, a single-stage-to-orbit (SSTO). The goal is to reuse vehicle components and eliminate multi-stage rocketry. To this end, Langley is assisting Lockheed Martin and Marshall Space Flight Center in developing and testing the X-33, an SSTO prototype slated to fly in 1999.

If America's 21st century space program is to be bold, advances in building structures in space are necessary. Need has evolved for affordable, higher precision structures that can keep sensors accurately trained on targets.

Langley is fabricating high-precision deployable structures that remain fixed and steady within an accuracy of four millionths of a inch. Next generation gamma ray telescopes, as well as other types of telescopes to search for, find and study Earth-like planets circling distant

stars demand such precision. Space structures are a springboard for many near- and deep-space missions, and Langley is part of this exciting challenge.

A NASA pilot makes a "windowless landing" aboard a NASA 737 research aircraft in flight tests aimed at developing technology for a future supersonic airliner. Cameras in the nose of the airplane relay images to a computer screen in the otherwise blind research cockpit.

Concepts of the Space Station in the early 1970s could not predict the advances in building structures in space in the late 1990s. Langley is currently fabricating high-precision deployable structures that remain fixed and steady within an accuracy of four millionths of a inch.

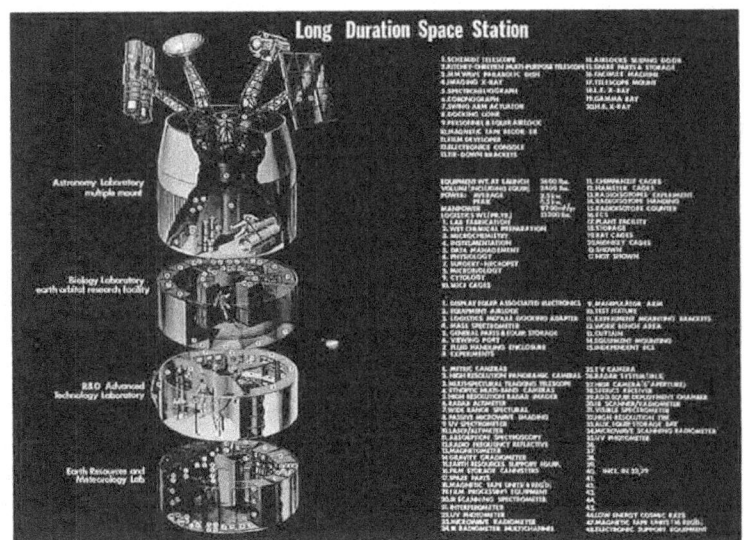

Lewis Research Center

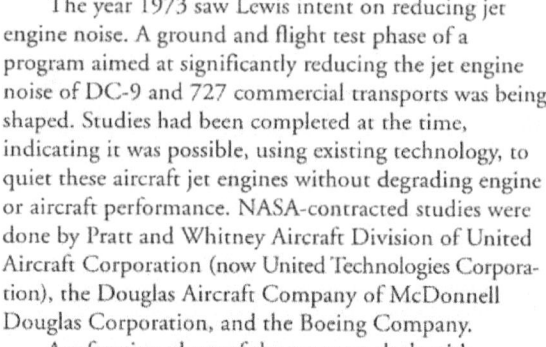

ocated in Cleveland, Ohio, Lewis Research Center is engaged in advancing propulsion technology, permitting aircraft to fly faster, farther and higher, and focusing its research talents on aircraft fuel economy, noise abatement, reliability, and reduced pollution.

Lewis is NASA's Center of Excellence in turbomachinery and commands top priority research in aeropropulsion. The Lewis Aeronautics Directorate and Aeropropulsion Research Program Office lead efforts in subsonic and supersonic propulsion, propulsion materials and structures, associated propulsion support technologies, as well as hybrid hyperspeed propulsion for aerospace application.

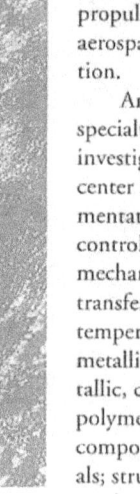

Among the specialty areas of investigation at the center are instrumentation and controls; fluid mechanics and heat transfer; high temperature metallic, intermetallic, ceramic, polymeric, and composite materials; structural mechanics and durability; basic chemistry and electrochemistry; tribology; photovoltaics; and microwave electronics. Both theoretical and applied research are pursued, the latter ranging from small laboratory experiments to full-scale component and systems tests.

Organized in 1941 by the National Advisory Committee for Aeronautics (NACA), the forerunner organization to NASA, this center is one of four assigned to NASA's Aeronautics and Space Transportation Technology Enterprise.

Research efforts at Lewis have affected every United States aircraft built since the early 1940s and have contributed to American leadership in international aviation. This is characterized by the Icing Research Tunnel, the world's largest, established in 1944. Research in this facility led to the development of ice protection technologies that largely have diminished icing problems for aircraft, although improvements in this arena continue at the tunnel.

NASA's Advanced Communications Technology Satellite (ACTS) is the technology leader of satellite communications, using futuristic dynamic hopping beams and advanced on-board switching and processing.

The year 1973 saw Lewis intent on reducing jet engine noise. A ground and flight test phase of a program aimed at significantly reducing the jet engine noise of DC-9 and 727 commercial transports was being shaped. Studies had been completed at the time, indicating it was possible, using existing technology, to quiet these aircraft jet engines without degrading engine or aircraft performance. NASA-contracted studies were done by Pratt and Whitney Aircraft Division of United Aircraft Corporation (now United Technologies Corporation), the Douglas Aircraft Company of McDonnell Douglas Corporation, and the Boeing Company.

A refanning phase of the program dealt with modifying a JT8D jet engine. Where the engine's present two-stage fan was contained, a larger, single stage fan would be installed. As a result of the larger single fan, the jet engine exhaust velocity was reduced, thereby significantly reducing jet engine noise. Acoustic treatment to muffle fan noise was incorporated in the engine housing. Two booster stages were added to the jet engine, lowering pressure compression to maintain the proper airflow conditions to the engine core.

Intensive work at Lewis was also carried out in 1973 on the guidance equations for NASA's liquid-hydrogen fueled Centaur launch vehicle. The Centaur, an upper stage for the Atlas and Titan rockets which was pioneered and perfected under Lewis management, had been updated to automatically check by computer the vehicle systems prior to and during the flight. A 16,000-word capacity computer had replaced the original 4,800 word capacity computer.

On the environmental front, Lewis assisted the Environmental Protection Agency in developing the technology for a low pollution automobile using a gas turbine engine. Still another move by Lewis on modifying automobile engine exhaust made use of technology originally designed to protect nuclear rocket nozzles from vibration.

That same year, Lewis engineers were also occupied by installing a solar system at the center. Not quite on the planetary scale, this experimental solar system collected sunlight to heat and cool a one-story office building. The hardware itself was being planned for Langley Research Center in Hampton, Virginia.

This rich history and diversity of work continues today at Lewis. Aeronautical and related programs in 24 major facilities and over 500 specialized research facilities are done at the center's 350-acre Cleveland location and 6,400-acre Plum Brook Station site in Sandusky.

One of those facilities, the Hypersonic Wind Tunnel, is the only nonvitiated hypersonic tunnel in the

United States. It produces high mass-flow rates of high temperature, uncontaminated air for simulating Mach 5 through Mach 7 velocities. Other wind tunnels support supersonic, transonic, and subsonic research.

Fundamental microgravity science experiments are tasked to the Lewis Microgravity Division. Academic, industry and government researchers have access to the Lewis ground-based research facilities. The program also supports the conduct of microgravity science experiments on the Shuttle and in the future aboard the International Space Station.

Combustion Module-1 (CM-1), for example, is the largest payload of Lewis-developed experiments ever flown. Slated to fly in 1997, lessons learned from CM-1 will improve awareness regarding the process of combustion. This appreciation may lead to practical applications both in space and here on Earth, such as improved fire safety and more efficient, clean-burning combustion engines. Moreover, CM-1 is a prototype for a permanent combustion experiment facility on board the International Space Station.

Plum Brook Station at Lewis received a taste of Mars. The Station hosted tests of the air bag landing gear that provided the Mars Pathfinder with a soft, upright landing when it encounters the rugged terrain of Mars in July 1997. In addition, Pathfinder was fitted with three experiments designed and built by Lewis to determine the effects of the Martian environment on surface exploration systems.

Back down on Earth, satellite communications is another Lewis specialty. The center manages the Advanced Communications Technology Satellite (ACTS) project, essentially an orbiting technology laboratory for communications. ACTS is examining and verifying advanced high-gain spot-beam Ka-band technologies. Industry, government, and university organizations are using ACTS to conduct a wide variety of integrated video, data, voice, and multimedia tasks. Experiment operations are now conducted 24 hours a day, seven days a week after it was deployed from a Space Shuttle in September 1993, then boosted into geostationary orbit.

Satellite operations and experiments are expected to continue to the year 2000. The 14 Federal Communications Commission filings for orbital slot locations demonstrate the impact that the ACTS experiment has had on the future of communications services.

Lewis Research Center is committed to playing a critical role in sustained aerospace leadership of the United States in the 21st century. It has the necessary technical assets, core capabilities, human capital, and facilities to support aeropropulsion and space research, technology, and development. Just as important, it has the support of its customers, stakeholders, and strategic partners in the business, university and public sectors. That confab assures that Lewis will have a major impact on the scope and direction of the nation's aerospace program and on national competitiveness in the approaching new millennium.

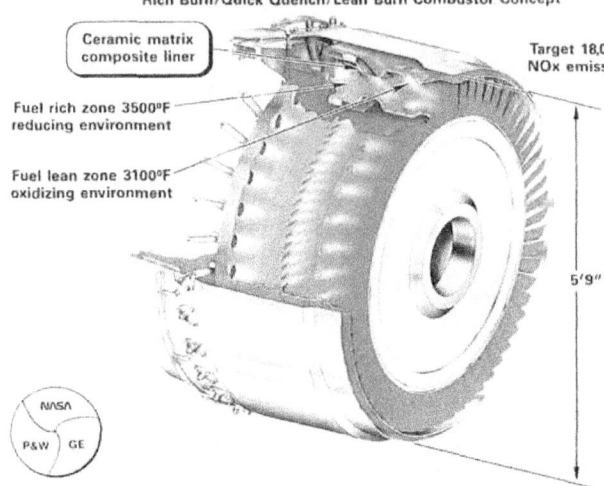

HIGH SPEED CIVIL TRANSPORT
Enabling Propulsion Materials Program

Rich Burn/Quick Quench/Lean Burn Combustor Concept

Ceramic matrix composite liner

Target 18,000 hours life
NOx emission index <5

Fuel rich zone 3500°F reducing environment

Fuel lean zone 3100°F oxidizing environment

5'9"

NASA
P&W GE

Advanced high-temperature materials with lifetimes up to 18,000 hours are critical to the design of low-emission powerplants for a future high-speed civil transport. NASA's Enabling Propulsion Materials program seeks to meet the technical challenges of supersonic airliner engine requirements.

Marshall Space Flight Center

The roar of rocketry has been part of George C. Marshall Space Flight Center's historic past, as it is in 1997. Space propulsion is the designated Center of Excellence area for Marshall, dedicated to supporting transportation systems development and microgravity research as mission areas.

Marshall is a multi-project management, scientific and engineering research and development establishment. Such epic projects as building the giant Saturn V booster and developing the main propulsion system for the Space Shuttle took place at this center.

Located on 1,800 acres inside the U.S. Army's Redstone Arsenal at Huntsville, Alabama, Marshall was officially dedicated by President Dwight D. Eisenhower in July 1960, by the transfer to NASA of part of the Army Ballistic Missile Agency.

The center's first director was Wernher von Braun, the renowned German rocket scientist who spearheaded the development of the boosters that kept the United States a competitor in the often-called "Space Race" with the former Soviet Union. Highlights of those early years included building the Mercury Redstone rocket that shot America's first astronaut into space, Alan Shepard, on a 15-minute test flight.

Von Braun's leadership ultimately propelled the United States to lunar distance. Riding atop the towering Saturn V rocket, Apollo astronauts became the first humans to step onto another world beyond Earth—the Moon.

Following those first footfalls on the Moon and subsequent Apollo Moon landings, Marshall was essential in making use of Apollo hardware and remaining Saturn boosters to create the Skylab program. In May 1973, America's first space station—Skylab—was boosted into Earth orbit by a Saturn V launcher. Skylab liftoff marked the end of an era in the history of the Marshall-developed Saturn V launch vehicle.

Several other significant events took place at Marshall in 1973, including development of the Space Shuttle. The Center was responsible for design and development of three major elements of the Shuttle system—the three main engines that power the orbiter, the external tank that holds the propellants for the engines during launch and ascent to orbit, and the solid rocket propellant strap-on boosters. In conjunction with the Shuttle, Marshall managed the Spacelab program in concert with Europe.

Marshall is responsible for the Michoud Assembly Facility in New Orleans, Louisiana. At this site, the Shuttle's external tank is manufactured. The center also is responsible for the assembly and refurbishment of the Shuttle's solid rocket motors that takes place at the Kennedy Space Center in Florida.

Marshall is moving forward with several enhancements for the Space Shuttle system. These will increase the performance of the Shuttle. Such enhancements include a new superlightweight external tank and a new high pressure oxidizer and fuel turbopump for the Space Shuttle's main engines.

Throughout its history, Marshall has managed many significant projects including Apollo's Lunar Roving Vehicle, the Hubble Space Telescope, over two dozen Spacelab missions since 1983, and the Advanced X-ray Astrophysics Facility (AXAF).

Marshall has been a leader in the development of scientific payloads and experiments flown aboard the Space Shuttle. Many of these payloads, including a variety of microgravity and astronomical experiments, have been carried within Spacelab, the reusable, modular research facility carried in the Shuttle's cargo bay. The center also operates NASA's Spacelab Mission Operations Control Center, a state-of-the-art facility from

In 1973, an engineer set controls as a second engineer uses Ergometer in the full-scale mockup of the Orbital Workshop at Marshall Space Flight Center. The workshop is the largest component of Skylab.

which all NASA Spacelab missions have been controlled. Marshall's control center can help with investigations undertaken by Shuttle astronauts, replan investigations if necessary, and monitor the overall health of scientific hardware.

Marshall is providing testing, manufacturing and assembly support for the International Space Station. That support includes developing the first major experiment facility for the huge orbiting complex, the Space Station Furnace Facility. Microgravity materials science research will be conducted in this facility.

As NASA's lead center for transportation system development and the Agency's Center of Excellence in propulsion, Marshall is, once again, at the forefront of rocket technology. The center manages NASA's Reusable Launch Vehicle (RLV) effort. An end product of this program is demonstrating a reusable launch system. New advanced technologies are to dramatically increase reliability and lower the cost of putting a pound of payload into space from $10,000 to $1,000.

One of three RLV projects managed by Marshall included four flight tests of the subsonic DC-XA, or Clipper Graham. Lifting off from a site in White Sands, New Mexico, the DC-XA flight tested advanced technologies such as lightweight composite propellant tanks, fuel lines and valves.

A second RLV program is the X-34, a small, reusable technology demonstrator vehicle. The fast-track X-34 program calls for demonstrating a vehicle that flies at eight times the speed of sound and reaches an altitude of 250,000 feet. The vehicle is to showcase low-cost reusable technology, autonomous landing, subsonic flights through rain, safe abort conditions, and landing in 20-knot cross winds. Marshall is providing design and development of the vehicle's main propulsion system.

The third RLV effort now in progress is the X-33, a craft designed to rocket to Mach 15. Built in partnership with Lockheed Martin, the X-33 integrates and tests advanced components and technologies necessary for industry to build a full-scale reusable launch vehicle. Lockheed Martin will design, build and conduct the first flight of the sub-scale X-33 test vehicle by March 1999, conducting at least 15 flights by December 1999.

The RLV program is a radical departure from the way NASA has done its rocket business in the past. While NASA is to develop the high risk technologies that industry cannot afford, the follow-on to the X-33, called VentureStar, is to be built by industry. NASA would use the full-scale VentureStar, not operate the launcher.

In accordance with NASA's goals to search for an understanding of the universe and explore the solar system, it is Marshall's vision to be the world leader in space transportation. The thrust of the Advanced Space Transportation Program is to focus on a broad spectrum of technological advances with the potential to reduce costs beyond RLV goals.

A low-cost booster technology project, called Bantam, is part of Marshall's work-in-progress. Investments are being made in innovative technologies for low-cost manufacturing and systems engineering of this booster. Air breathing rockets are part of the advanced space transportation work at Marshall, as well. These systems would use atmospheric oxygen for its oxidizer thus eliminating the need to carry stored oxygen in the rocket.

While the rumble from huge Saturn V rockets has fallen silent, Marshall Space Flight Center remains as NASA's primary center for bringing ideas, innovation and new technologies forward to dramatically increase reliability and lower the costs of putting payloads, people, and aspirations into space.

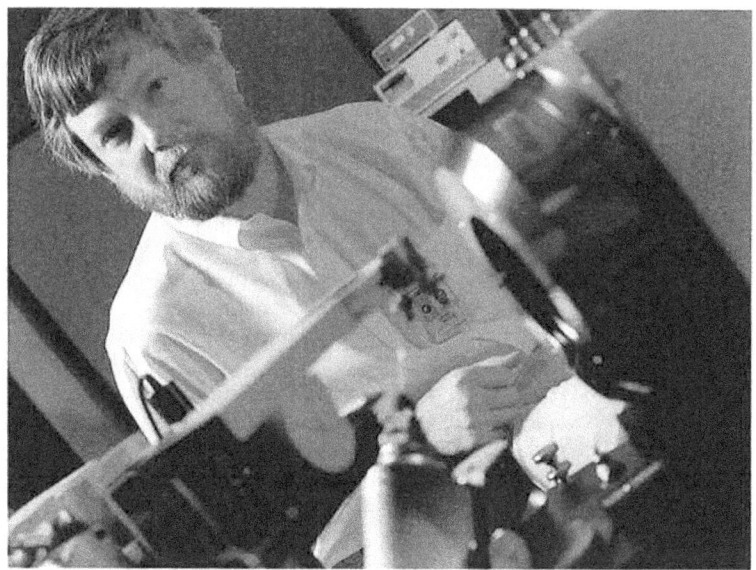

An investigator at the Biophysics Research Office at Marshall Space Flight Center prepares the High Brilliance X-ray instrument for a test run. The instrument is used for protein crystal research.

John C. Stennis Space Center

NASA's premier center for testing large rocket propulsion systems for the Space Shuttle and future generations of launch vehicles is the John C. Stennis Space Center. Close to 14,000 acres make up the operations complex, placed near the Mississippi Gulf Coast. Stennis is NASA's lead center for rocket propulsion testing and for commercial remote sensing.

Formerly named the Mississippi Test Facility (MTF), the center was tasked from 1966 to 1970 to test fire all first and second stages of the Saturn V rocket used in the Apollo manned lunar landing and Skylab programs. The Saturn first stage contained five F-1 engines that together yielded a ground-shaking wallop of 7.5 million pounds of thrust.

The end of the Saturn series also meant an end to the Mississippi Test Facility's primary support role. But in March 1970, then President Richard Nixon approved development of a reusable shuttle vehicle that could be launched vertically like the Saturn V, but would return to Earth and land on a runway like an airplane. The Space Shuttle was tagged as the world's first and only reusable launch vehicle.

It was declared in April 1972 that MTF would take on the role of testing the main engines for the Space Shuttle. As the center's new roles and missions gained momentum, the MTF became a hub of activity by 1973 in support of the Space Shuttle.

In June 1974, MTF became the National Space Technology Laboratories. From 1975 onward, flight acceptance testing of the Space Shuttle Main Engine (SSME) has been an ongoing assignment. The data accumulated from these ground tests, which simulate flight profiles, are analyzed to ensure that engine performance is acceptable and that the required thrust will be delivered in the critical moments of Shuttle ascent.

Every SSME must undergo certification and acceptance testing at Stennis prior to flight. The engines sit vertically in one of the three large test stands where they undergo a series of test firings. Once proven flight worthy, Stennis ships the engines to Kennedy Space Center, Florida, for installation on a Shuttle orbiter. A new phase of SSME testing recently began, focused on Block I and Block II engines. It is anticipated these will be more reliable and less expensive to operate.

Propulsion systems research and development at Stennis has an "applied research" orientation. Some facilities of particular importance are the Diagnostics Testbed Facility used for exhaust plume diagnostic sensor development and evaluation; the E-1 Facility used for the development and testing of individual engine components; and the E-2 Facility used for thermal testing of materials, such as cryogenic fuel tanks for hypersonic aircraft of the future.

Now under the name of John C. Stennis Space Center, the work ahead builds on a 30 year past of rocket engine testing. Granted its new lead center chores for rocket propulsion testing in May 1996, Stennis is currently responsible for managing all of NASA rocket propulsion test assets, activities and resources. Other responsibilities include developing testing and facility investments, consolidating strategies, as well as determining where tests will be performed across NASA centers.

The first stage of the powerful Saturn V rocket is hoisted by crane for inspection at the Mississippi Test Facility, now John C. Stennis Space Center. The facility was responsible for testing all first and second stages of the rocket during the Apollo program.

Two programs that will pave the way for future space travel are the Reusable Launch Vehicle (RLV) and the Evolved Expendable Launch Vehicle (EELV) programs.

The RLV program is a joint NASA-industry partnership with the goal of developing a new generation of rockets expected to radically cut the cost of placing payloads into space. Stennis will conduct or manage most RLV propulsion testing. NASA's selection of Lockheed Martin Skunk Works to build and fly the X-33 test vehicle means development testing of its engine has been earmarked for Stennis. Specifically, the X-33 cryogenic fuel tanks and components for a new half-scale Rocketdyne aerospike engine are being prepared for Stennis testing and evaluation. That smaller-scale aerospike engine will ultimately lead to a full-scale version on Lockheed Martin's commercial space plane, the VentureStar.

A top priority for the Air Force is establishment of an Evolved Expendable Launch Vehicle (EELV). Modern rocket technology is called for to design and build a family of expendable, or non-reusable, rockets to replace the aging Delta, Atlas and Titan boosters.

Stennis has played important roles in demonstration tasks, engine and system testing for the program. The EELV will deploy payloads weighing from about 2,500 pounds to 45,000 pounds into a low-Earth orbit. The goal is to reduce the cost of launch vehicles from 25 percent to 50 percent. Stennis Space Center's test capabilities will allow the Air Force and its industrial contractors to use existing test facilities.

The goal of NASA's Commercial Remote Sensing Program (CRSP) is to enhance U.S. competitiveness through the use of remote sensing, Geographic Information Systems and related technologies. The CRSP Office at Stennis administers several partnership programs designed to share NASA's technology and expertise with U.S. industry.

Stennis provides the bridge between NASA's Small Satellite Technology Program and the private sector for developing commercial remote sensing applications. Projects include preserving the tropical rain forest in Central America, studying sea surface temperatures to determine conditions for red tide outbreak, analyzing plant stress and monitoring cultural and historical archaeological sites.

The purpose of NASA's Earth Systems Science Office at Stennis is to develop an understanding of key biological, chemical, geological, and physical processes and man's influence on these processes. The research focus is the study of coastal processes in support of NASA's Mission To Planet Earth program.

Stennis work in remote sensing has proven valuable in promoting the commercial utility of space for Earth observations technology. Remote sensing and Geographic Information System programs are being appraised at the center, determining their economic benefit, and how best to transfer such approaches to the private sector.

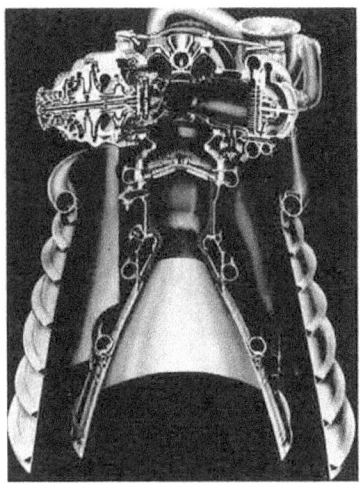

The Space Shuttle was originally designed as a manned reusable space vehicle which would carry out various space missions in Earth orbit. Stennis will conduct or manage most propulsion testing.

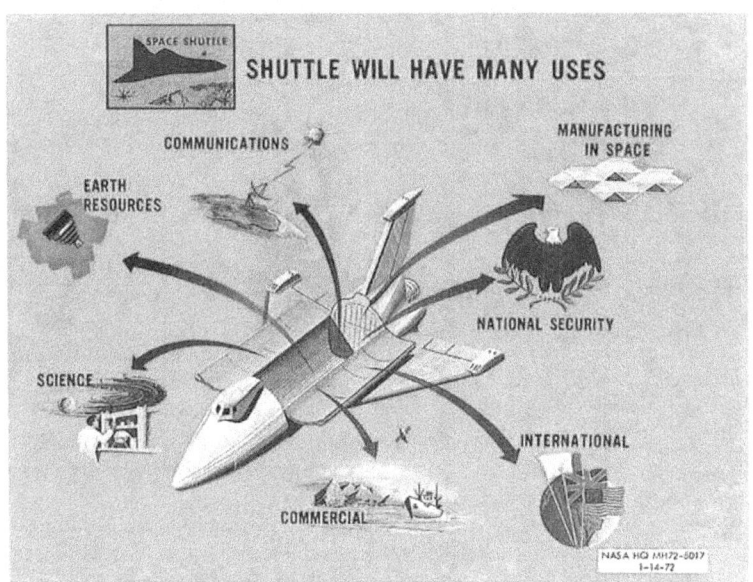

Designs in the early 1970s for Space Shuttle Main Engines (SSME) depict plans to use liquid oxygen/liquid hydrogen for both the orbiter and booster engines. Every SSME must undergo certification and acceptance testing at Stennis prior to flight.

Jet Propulsion Laboratory

S lipping across the vacuum void of interplanetary space, U.S. spacecraft have made up-close-and-personal treks to all the known planets except Pluto. The names of a few of these far-flung robot explorers are likely to sound familiar: Mariner, Viking, Voyager, Magellan, and Galileo. Each spacecraft has helped unravel the secrets of the solar system, adding to our zeal for exploration and fostering a sense of wonder about the surrounding universe.

Nestled within the San Gabriel Mountains near Pasadena, California, the Jet Propulsion Laboratory (JPL) is the leader of NASA's Space Science Enterprise. Managed for NASA by the California Institute of Technology, JPL is the space agency's Center of Excellence in deep space systems.

JPL also manages the worldwide Deep Space Network, radio dish complexes situated in the Mojave Desert near Goldstone; near Madrid, Spain; and near Canberra, Australia. This network communicates with spacecraft and conducts scientific investigations.

Looking back to 1973, the laboratory was undertaking several major flight projects, in addition to preparing for two major missions to come. Mariner 9 became the first spacecraft to orbit another planet—Mars. Studies of the data acquired by the Mariner continued throughout 1973, offering new discoveries about the Mars atmosphere, topography, and gravity field.

Yet another Mariner was leaving its launch pad on November 3, 1973. It would later become NASA's first dual-planet mission, flying by Venus February 5, 1974, then onward to make a succession of Mercury flybys in 1974 and 1975. This spacecraft, Mariner 10, was the first to use "gravity-assist"—a method that is akin to celestial square-dancing, using the gravity of one planet to slingshot a spacecraft onward to another. The innovative technique has been applied to many spacecraft missions over the years, with more being planned.

Today, the Jet Propulsion Laboratory's agenda is a far-reaching one compared to its early, pioneering work.

JPL has long been a reservoir of the best scientific minds and engineering muscle to conceive and execute robotic exploration of the Moon and planets. Ranger and Surveyor missions to the Moon prepared the way for Apollo astronauts to explore that barren and crater-pocked globe. Mariner spacecraft reached Mercury, Venus and Mars, unlocking the mysteries of those distant worlds. Each successful mission was a trailblazer.

In the mid-1970s, strange shadows fell across the face of Mars. Two Viking robot landers touched down on the reddish Martian sands in July and September 1976. As two Viking Orbiters circled the red planet, the Viking Lander duo carried out an extensive reconnaissance of Mars, including a search for evidence of Martian life. The elaborate Viking program involved several NASA centers, with JPL building the Viking Orbiters, conducting mission communications, and eventually assuming responsibility for mission management.

Launched in 1977, JPL's twin spacecraft Voyager 1 and Voyager 2 chalked up interplanetary mileage by flying past Jupiter and Saturn. Voyager 2 then went on to encounter the Planet Uranus in 1986 and Neptune in 1989. These Voyagers are just that, both voyaging out toward interstellar space, still communicating scientific findings, perhaps until the second decade of the 21st century.

Ulysses is a joint NASA-European Space Agency undertaking to explore, for the first time ever, the Sun's polar regions. JPL scientists are analyzing data gathered by Ulysses in 1994 and 1995, preparing for yet another set of polar fly-overs in 2000 and 2001.

In 1996 into early 1997, images and data relayed from Galileo as it circled the massive planet were staggering in detail. Particularly exciting are stunning photos taken of Jupiter's moon, Europa. Scientists wonder whether or not this ice-covered moon might harbor primitive life. Galileo's tour-of-duty has been so spectacular, NASA is extending its data-gathering tasks into 1999.

JPL has designed and built the Cassini mission to Saturn, slated for launch in 1997. It arrives at the ringed world in 2004. Cassini carries the European Space Agency's Huygens probe, destined to land on the surface

The Cassini spacecraft is placed on a trailer at NASA's Jet Propulsion Laboratory for transport to test facilities.

of Titan, Saturn's largest moon. This moon is an apparent locale for organic chemistry possibly like that which led to life on Earth.

Mars once again is being targeted. The planet is to be orbited starting in 1997 by the Mars Global Surveyor while the Mars Pathfinder is built for landing. After touchdown, the Pathfinder lander will dispatch Sojourner, a micro-rover.

JPL's Surveyor program for Mars involves lander and orbiter missions with each launch opportunity— about every two years—well into the 21st century. Work is underway to detail a 2005 robotic return sample mission from Mars. This string of Surveyor orbiter and landers may well answer conclusively whether or not life has been, or is even now, present on Mars.

Other JPL missions being readied include Stardust. To be rocketed spaceward in 1999, this probe heads for comet Wild-2, collecting dust and other materials tossed off from the object, then return those samples to Earth. Actually landing on a comet and hauling back to Earth cometary material is the 21st century flight of JPL's Champollion spacecraft.

The innovative New Millennium program is a high-tech initiative that promises to overhaul how space science missions are conducted. These craft are to be smaller, more compact and more versatile than the robotic explorers of the past. This breed of microspacecraft is geared to validate state-of-the-art electronics, propulsion, guidance and control, including a host of sensors and other devices. New Millennium also spearheads new industry and university partnerships with JPL.

New Millennium's Deep Space-1, being readied for launch in 1998, is to be propelled by an ion-engine, rendezvousing with both a comet and asteroid. Deep Space-2 centers on New Millennium Mars Microprobes, small penetrators that will slam into the Martian surface, dispatched from the Mars Surveyor 98 spacecraft. Developing the technology required to search for planets circling other stars is New Millennium's Deep Space-3 interferometry mission in 2000. Deep Space-4 is to test an array of Earth remote sensing technologies, also in 2000.

Another 21st century mission is Pluto Express, a two-probe encounter with the outermost planet now known and its moon, Charon, perhaps in 2010.

JPL is preparing the turn-of-the-century Space Infrared Telescope Facility (SIRTF) project. This 2001 spacecraft is being blueprinted to probe hundreds of thousands of celestial objects invisible to conventional telescopes. SIRTF would complete NASA's series of

"Great Observatories," telescopes based in space that study the universe at wavelengths ranging from visible light to x-rays and gamma rays. Part of SIRTF's work is to probe the realm of very young galaxies in the process of forming.

Although known world-wide for space exploration, JPL contributes to Earth observation missions as well. In fact, JPL is the largest provider of science instruments for NASA's Mission to Planet Earth program.

JPL's sensor technology expertise has been applied to the U.S.-French TOPEX/Poseidon, lofted in 1992 to acquire data on the oceans' role in climate change. The Spaceborne Imaging Radar—C/X-band Synthetic Aperture Radar—has collected information on geological processes and air-sea interactions and is a collaborative project of NASA, the German Space Agency and the Italian Space Agency.

Japan's Advanced Earth Observing Satellite (ADEOS), orbited in August 1996, carries a NASA Scatterometer (NSCAT) to help decipher change in Earth's climate, and to help improve accuracy of weather forecasting and the nature of destructive storms. In 1997, JPL selected four industry teams to implement LightSAR, a proposed new Earth-imaging satellite. LightSAR would use advanced technologies to reduce the cost and enhance the quality of radar-based information for scientific research, commercial remote-sensing and emergency management applications.

Mariner 10, launched on November 3, 1973, was the first to use "gravity-assist," using the gravity of one planet to slingshot a spacecraft onward to another. The innovative technique has been applied to many spacecraft missions over the years, with more being planned.

Technology Transfer and Commercialization

NASA has traditionally measured its progress in terms of technical performance, cost and schedule. Now, in the post Cold War era there is another measure: contribution of technology to national economic security.
—*Agenda for Change*

Because they are challenging and demanding, NASA programs generate a great wealth of advanced technology. This bank of technology is a national asset that can be reused to develop commercial products and processes, which benefit the U.S. economy by creating companies and jobs, and as a result contribute to the Gross Domestic Product. According to NASA Associate Administrator Robert E. Whitehead, NASA has helped generate almost one million high-quality jobs, more than $40 billion in annual exports and almost $30 billion in positive balance of trade.

Through its Technology Transfer Program, NASA employs a variety of mechanisms to stimulate the transfer of aerospace technology to other sectors of the economy. The program is managed by the Commercial Development and Technology Transfer Division of NASA's Office of Aeronautics and Space Transportation Technology headquartered in Washington, D.C. Recently reorganized to include elements of the Office of Space Access and Technology, the Office of Aeronautics and Space Transportation Technology functions as a strategic alliance between NASA's aeronautics and space programs for developing then transferring and commercializing technologies.

Another new aspect of the Office of Aeronautics and Space Transportation Technology is the addition of the Reusable Launch Vehicle program; in partnership with the Air Force and private industry, NASA is developing and demonstrating new technologies for the next generation of reusable space transportation. This will ultimately reduce the cost of access to space, and thereby create a new market: the commercialization of space.

The Commercial Development and Technology Transfer Division coordinates the activities of the NASA Commercial Technology Network (CTN) throughout the United States. The CTN includes a commercial technology office based at each of NASA's 10 field centers. The offices serve as regional managers of the Technology Transfer Program, promoting and facilitating transfer and commercialization of technology that has significant potential for secondary use.

Since *NASA Commercial Technology: Agenda for Change* was written in 1994, NASA has been implementing a new way of doing business through new practices. The Commercial Technology Program has been inviting industry to define and lead joint R&D projects that are relevant to the NASA mission but not driven by specific project requirements, with industry sharing the costs. Dual-use technology, where the efforts can be applied to mission objectives and industry commercial goals; a push to acquire commercial off-the-shelf technology to meet program requirements; targeting small business in outreach and information exchange efforts; and establishing local, state and regional alliances to leverage resources and build individual technology networks are a few other examples of NASA's new approach.

The following pages contain a summary of the mechanisms employed by NASA to promote and facilitate technology transfer, including the structure of the CTN, the types of assistance provided by the Regional Technology Transfer Centers and affiliated organizations, and the technology transfer activities of NASA's field centers. To illustrate some of the details of today's technology transfer work, the field center summary focuses on activities within one organization—Langley Research Center's Technology Applications Group.

Spotlight on Langley Research Center

Representative of the Technology Transfer & Commercialization Offices is Langley Research Center's Technology Applications Group (TAG), located in Hampton, Virginia. TAG's primary goal is to encourage broader utilization of Langley-developed technologies in the American industrial community. According to TAG director Dr. Joseph Heyman, the group is organized with very little internal structure to give it the maximum flexibility to meet the requirements of a very dynamic marketplace.

Dr. Heyman says that key elements of TAG can be illustrated as four spokes of a wheel, the first being information and data. In an effort to improve internal communications, data archiving, and to develop databases and other information for the center and its customers, TAG is developing better ways to use electronic media such as disseminating information on technologies available for transfer through the World Wide Web.

The second element of TAG is the technology patent structure, which handles intellectual property such as hardware patents, software copyrights and technical know-how. The third is the function as a financial resource for small businesses, including the Small Business Innovation Research program (SBIR), the Small Business Technology Transfer program (STTR) and the Aerospace Industry Technology Program (AITP). The purpose of these programs is to stimulate businesses to partner with the government and work in focused mission areas that have dual use.

The fourth spoke is the technology transfer team—engineers and scientists who foster technology partnerships with industry. "They discover business needs by communicating with business; they identify the Langley technologies by knowing the center, not just by reading papers but by knowing the people," said Heyman. The team advertises Langley-developed technologies available for transfer through the Internet, *NASA Tech Briefs* and other journals,

This birds-eye view of NASA Langley Research Center was taken with an extreme wide-angle lens to capture most of its approximately 800 acres. The center's Technology Applications Group facilitates the transfer of Langley technology to the commercial sector.

An airplane model is tested in Langley's National Transonic Facility, one of Langley's many wind tunnels.

The center's efforts have produced measurable success: Langley's technology transfer and commercialization efforts in Fiscal Year 1996 resulted in three STTR and 50 SBIR Phase I Awards, five STTR and 21 SBIR Phase II Awards, ten licenses, 49 non-aerospace and 24 aerospace Memoranda of Agreement/Understanding, and 125 software releases. The parawing, the precursor to the hang glider developed at Langley in the late 1950's, was inducted into the U.S. Space Foundation's Hall of Fame in 1995. Langley was a recipient of NASA's 1996 Software of the Year award with the Tetrahedral Unstructured Software System (TetrUSS), an aerodynamic analysis and design system widely used throughout U.S. industry, government and universities to study aerodynamics and other problems.

By taking the initiative to know its customers and understanding their requirements, identifying new technologies for transfer and forming partnerships with industry, TAG continues to facilitate the transfer of Langley technology to the commercial sector. And, as Dr. Heyman said, "We are entering a new era of greater partnership with industry and recognizing that enhancing U.S. economic strength is part of our mission, part of our reason for being."

talks, and the Technology Opportunities Showcase (TOPS). TOPS was last held in 1995, providing a venue for industry to explore Langley partnership opportunities.

TAG provides the overall leadership for implementation of the center's Technology Transfer & Commercialization Program. This includes:

- leading the center's processes for early identification of technologies of high commercial potential;
- promoting the expedient transfer of new technologies to the commercial sector;
- achieving the non-aerospace uses of Langley technology by identifying potential technology applications and creating teams of non-aerospace customers and Langley technologists to accomplish the transfer process;
- coordinating the Langley program with appropriate NASA Headquarters offices, other NASA centers, and other government agencies; and
- supporting the technology transfer process for aerospace customers.

Langley was a recipient of NASA's 1996 Software of the Year award for the Tetrahedral Unstructured Software System (TetrUSS), an aerodynamic analysis and design system used here on the Boeing 747 to study aerodynamics and other problems.

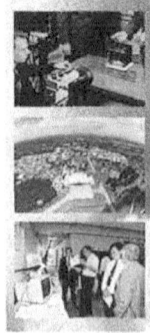

Langley Research Center Applications Projects

In addition to spinoff products, NASA's applications projects also represent successful technology transfer. These projects contribute to Langley's aerospace missions and also solve significant public sector or industrial problems through the redesign of existing NASA technology. Often, these projects also result in spinoff products, but their benefits to mankind begin much earlier.

For instance, Langley's High Performance Computing and Communications/Information Infrastructure Technology and Applications (HPCC/IITA) Program developed an Internet access system for K-12 schools that provides up to an 80% cost savings over standard connections. The commercialization outgrowth of this system is ATLAS (the Affordable Technology to Link America's Schools), a strategic plan to rapidly deploy the new connectivity design through the training of personnel, implementation of the network, and partnerships. The Virginia State HUD Office is negotiating with NASA to support the application of ATLAS to the Neighborhood Networks (NN) Project within the state, and Unified Research Laboratories (URLabs) of Virginia has signed a Memorandum of Understanding with Langley to support the ATLAS goal to offer affordable Internet connections to 70,000 K-12 school sites by the year 2000.

Another current Langley applications project involves several new medical device technologies resulting from the center's expertise in instrumentation design and development and nondestructive measurement science. These include devices for non-invasive evaluation of diaphragm function, diagnosis of pressure ulcers, non-invasive intracranial pressure measurement, circumferential pressure probe for urodynamics, and a CCD mosaic for digital mammography.

One device, for the evaluation of diaphragm function, utilizes an ultrasound system positioned to view the lateral aspect of the diaphragm and automates the selection of the internal and external views of the diaphragm recorded. By judging and comparing the thickness of the diaphragm, a healthcare professional can see if a ventilated patient is breathing on his or her own, since only active breathing produces variations in thickness. The healthcare professional can then judge the minimal pressure needed for a ventilator and gradually increase the patient's independence by gradually reducing the pressure.

A Langley technology that could make everything from audio speakers to heart pumps smaller and more efficient is THUNDER (Thin-Layer Composite-Unimorph Piezoelectric Driver and Sensor). Piezoelectric materials generate mechanical movement when

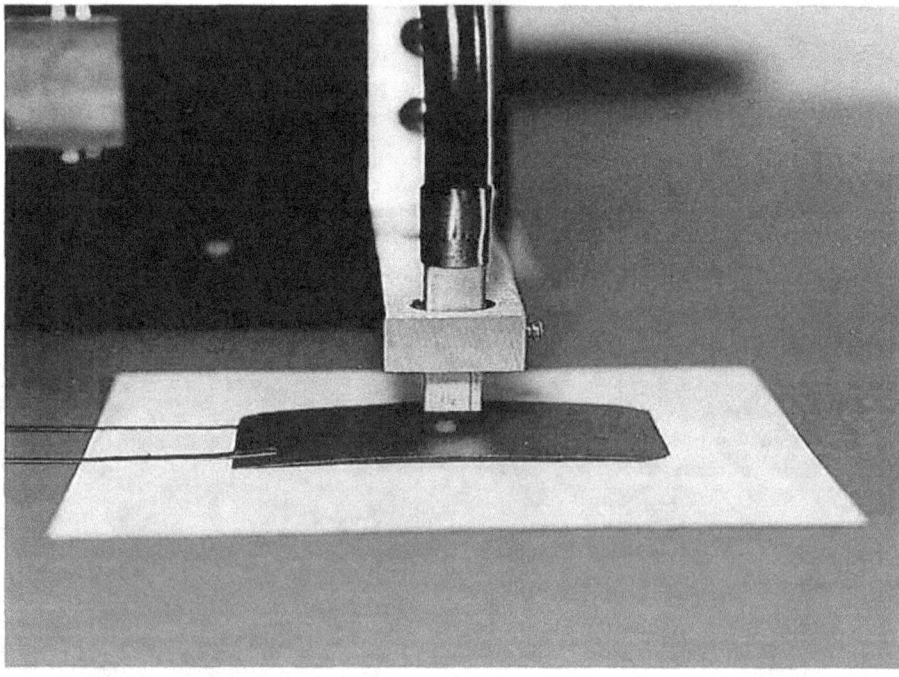

A fiber-optic displacement sensor measures the displacement of a THUNDER wafer. THUNDER is a NASA technology that could make everything from speakers to heart pumps smaller and more efficient.

subjected to an electric current and generate electrical charge in response to mechanical stress. Langley researchers developed a piezoelectric material that is superior in several ways to those that are currently commercially available. It is tougher, should allow lower voltage operation, has far greater displacement, has greater mechanical load capacity, can be easily produced at a relatively low cost, and lends itself well to mass production. For this technology, *Research and Development* magazine presented Langley in 1996 with an *R&D 100 Award*, which recognizes the innovators of the 100 most technologically significant new products of the year. Six companies have signed Memoranda of Agreement to commercially develop THUNDER technologies and more than 20 other companies are negotiating agreements.

Another applications project involved NASA, the FAA and industry partners, which together provided the technology for a revolutionary new system used during the 1996 Summer Olympic Games to move air traffic efficiently and safely in uncontrolled airspace. The Olympics offered a rare opportunity to demonstrate advanced communications/navigation/surveillance flight systems for future air traffic management and emergency response.

The technology behind the system combines the use of digital data link communications and Global Positioning System (GPS) satellite navigation technologies, which provide pilots information about the positions of other aircraft and ground-based GPS systems. Ground crews monitor aircraft positions and then relay the information to the pilots. This enables the specially equipped aircraft to perform self-dispatch operations in a "free flight" mode.

ARNAV Systems, Inc., a member of NASA's Advanced General Aviation Transport Experiment (AGATE), was one of the private companies involved in the Olympic demonstration. Their VHF GeoNet digital data link network was used to provide automatic dependent surveillance-broadcast capability for aircraft operating within FAA temporary flight restricted areas.

Application projects originate in various ways, including from requests for assistance from other government agencies or NASA technologists themselves.

ARNAV Systems, Inc. was one of the private companies involved in an Olympic demonstration of new air traffic control technologies. Their system was used to provide automatic dependent surveillance-broadcast capability for aircraft operating within FAA temporary flight restricted areas.

Technology Transfer Centers

The NASA Commercial Technology Program encompasses a national network of specialized centers and programs that assist U.S. businesses and industry in accessing, utilizing and commercializing NASA-funded research and technology. These organizations work closely with the NASA field centers to provide a full range of technology transfer and commercialization services and assistance. Within the program is

Employees of the Southern Technology Applications Center (STAC) at the University of Florida's College of Engineering in Alachua exchange ideas at their annual Strategic Planning meeting. From left to right are Bill Huffman, Debra Mills, Dave Sapuppo (standing), Jim Nicholson, Ellen Boukari and STAC director Ron Thornton. STAC is one of six NASA Regional Technology Transfer Centers.

the National Technology Transfer Network, consisting of the National Technology Transfer Center (NTTC) and six Regional Technology Transfer Centers (RTTCs).

The NTTC, located at Wheeling Jesuit College in Wheeling, West Virginia, is the hub of the network, linking U.S. companies with federal technological resources through its National Gateway. The Gateway provides businesses with rapid access to NASA and other federal technologies, expertise and facilities through various means including the World Wide Web at **http://www.nttc.edu** and discussions at trade shows.

NTTC also administers NASA's primary tool for managing its commercial resources: TechTracS (Technology Tracking System). TechTracS is an on-line network of databases located at each of the 10 field centers, Headquarters and NTTC and can be accessed on the World Wide Web at **http://ntas.techtracs.org**. It is designed to maintain a complete inventory of all NASA technologies, help manage the identification of new technologies and their commercial potential, administer intellectual property protection and seek commercialization partners.

The six RTTCs in the national network are:
Northeast (Connecticut, Maine, Massachusetts, New Hampshire, New Jersey, New York, Rhode Island, Vermont)

The Center for Technology Commercialization (CTC) in Westborough, Massachusetts, is a not-for-profit corporation that manages NASA's RTTC in the Northeast. With a focus on technologies and facilities available at Goddard Space Flight Center, CTC works with regional companies to obtain and commercialize technologies developed by federal laboratories, universities, and industry. CTC has eight Satellite Technology Transfer Centers located in the six New England states, New York and New Jersey. As of late 1996, CTC had established four new companies, secured 30 licenses, completed 59 partnership agreements, and provided services to 3,400 companies.

Far West (Alaska, Arizona, California, Hawaii, Idaho, Nevada, Washington)

The Far West Technology Transfer Center, located at the University of Southern California in Los Angeles, focuses on linking regional companies with NASA and federal laboratories to license technology and/or enter into cooperative development agreements. The center operates its unique Remote Information Search Service to generate information from hundreds of federal, state and local databases; staff works with companies to identify opportunities at specific federal laboratories and other resources.

Mid-Atlantic (Pennsylvania, West Virginia, Virginia, Delaware, Maryland)

The Mid-Atlantic Technology Applications Center (MTAC) in Pittsburgh, Pennsylvania, can utilize a variety of resources to provide clients with "virtual" R&D. With close associations with Langley Research Center and Goddard Space Flight Center, MTAC supplies support and guidance from initial inquiry to market. MTAC also has initiated Public Service Partnerships, which are technology commercialization activities designed to solve problems and create new products that improve quality of life; one example is a Memorandum of Understanding Langley Research Center signed with the Pittsburgh Bureau of Fire, facilitated by MTAC.

Southeast (Alabama, Georgia, Florida, Kentucky, Louisiana, Mississippi, North Carolina, South Carolina, Tennessee)

The Southern Technology Applications Center (STAC) at the University of Florida's College of Engineering in Alachua expedites technology transfer and economic development through affiliates in nine southeastern states. STAC also works closely with Marshall Space Flight Center, Kennedy Space Center, and Stennis Space Center as a partner in the NASA Southeast Technology Transfer Alliance. The Alliance leverages resources and promotes NASA technologies, expertise and facilities for industry's use.

Midwest (Illinois, Indiana, Michigan, Minnesota, Ohio, Wisconsin)

The Great Lakes Industrial Technology Center (GLITeC) in Cleveland, Ohio, has helped more than 2,100 companies in its six-state region with technology-based problem solving and commercialization assistance services. Less than a mile from Lewis Research Center, GLITeC has special access to Lewis technology and staff and has initiated several technology transfer programs with Lewis. One such program is the series *Technology Dialogue Over Lunch*, which GLITeC uses to create relationships between Lewis and Ohio industry. The RTTC also collaborates with the federal Technology Access for Product Innovation (TAP-IN) program, implementing its market campaigns. GLITeC is managed by Battelle Memorial Institute, a nonprofit independent research organization.

Mid-Continent (Montana, Wyoming, Utah, Colorado, New Mexico, Texas, Oklahoma, Kansas, Nebraska, Arkansas, North Dakota, South Dakota, Iowa, Missouri)

Headquartered at the Texas Engineering Extension Service (TEEX) of the Texas A&M University System in College Station, the Mid-Continent Technology Transfer Center (MCTTC) works within a 14-state area to match technology-related needs to solutions drawn from a nationwide pool of resources. By exercising its unique position in TEEX, MCTTC extends its technology transfer resources to other programs, including the Texas Manufacturing Assistance Center and the Economic Development Administration at Texas A&M. Reporting directly to Johnson Space Center, the MCTTC provides public and private affiliates with the tools, expertise and resources to grow and compete on a national level.

Celebrating the signing of the Memorandum of Understanding between the Mid-Atlantic Technology Applications Center (MTAC), NASA and the City of Pittsburgh Bureau of Fire (PBF) are NASA Administrator Daniel S. Goldin, MTAC Executive Director Lani S. Hummel, Business Development Specialist Robert Saba, and Robert Hirosky and Charlie Dickinson of the PBF.

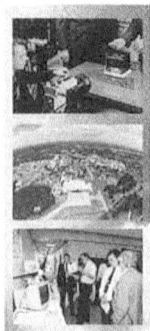

Affiliated Organizations and Programs

In addition to the Technology Transfer & Commercialization Offices, NTTC and RTTCs, the NASA Technology Transfer Network spans into several other NASA and affiliated organizations and programs. The trend is to get closer to the customer, becoming more accessible and proactive in technology transfer.

Technology Incubators

In the spirit of the new direct approach to commercial technology development, NASA established five technology incubators to house fledgling businesses during their vulnerable beginnings. For example, **Ames Technology Commercialization Center** (ATCC) at Ames Research Center house 25 high-technology start-up companies as of early 1997. All these companies have a specific relationship with NASA: they use NASA technologies to enhance their product or service, provide spin-in technologies to NASA to leverage existing research/engineering/administrative efforts, or collaborate with Ames on technical projects. Through licences, contracts and Space Act Agreements, the companies are utilizing NASA software, hardware, aerospace technology, robotics, and expertise. ATCC is operated by The Enterprise Network, which provides teams of volunteer executives to assist tenants.

The tenants of the **Mississippi Enterprise for Technology** at Stennis Space Center have combined sales exceeding $6 million. Two new incubators launched in July and August 1996 respectively expect similar success: the **Lewis Incubator for Technology** at Lewis Research Center, the result of a partnership with Enterprise Development, BP America, Inc., and the Great Lakes Industrial Technology Center; and the **Florida/NASA Business Incubation Center** at Kennedy Space Center, managed through a joint

NASA Lewis researcher Dr. Mark Bethea (right) demonstrates his Stereo Imaging Velocimetry technology to Lewis Incubator for Technology (LIFT) associates (standing, left to right) Kim Veris, NASA Lewis Commercial Technology Office; Wayne Zeman, LIFT executive director; and (kneeling) Rob Usher, Great Lakes Industrial Technology Center. This breakthrough technology has already improved commercial products from vacuum cleaners to steel production.

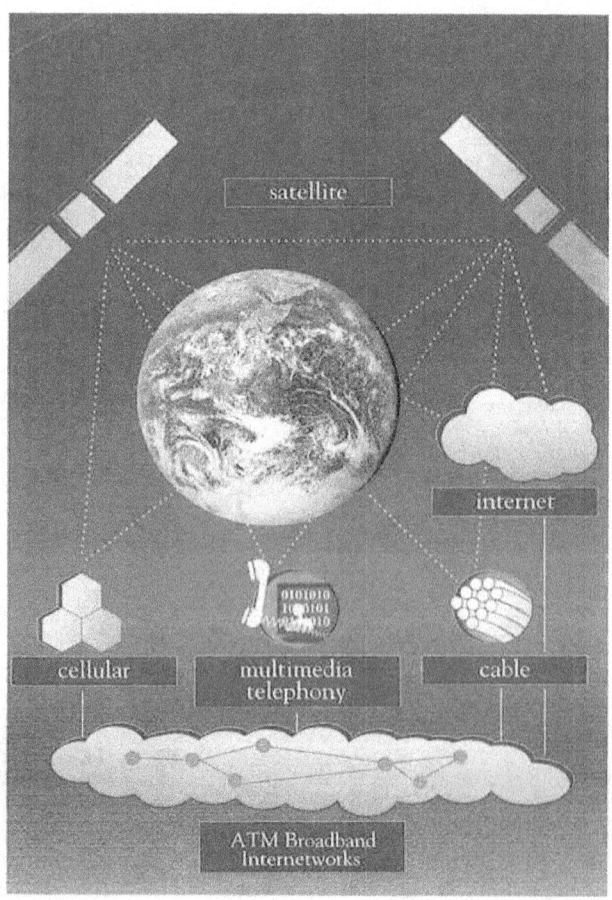

The Center for Satellite and Hybrid Communications Networks at the University of Maryland, established in 1991, specializes in communication networks that utilize both satellite and terrestrial resources. The diagram shows the relationship between the Earth, satellites and the various networks the Center employs.

partnership of Brevard Community College, the Florida Technological Research and Development Authority and Kennedy Space Center.

Partnerships

Various partnership options are available for industry to tap NASA resources, each with a different level of NASA participation and funding. For example, a company may enter a Reimbursable Space Act Agreement with NASA and essentially pay for a good or service; in a Nonreimbursable Space Act Agreement, both NASA and the company contribute resources to a project of interest to both parties. A Memorandum of Understanding involves no exchange of funds or resources, but is a statement of policy, practice or intention affecting a project of mutual interest.

Another partnership is the Small Business Innovation Research (SBIR) contract, which is a means for small businesses to participate in federal R&D activities and receive funding. NASA's program has produced hundreds of new systems that not only advance NASA's capabilities but many of which have become commercial products, over 60 of which have been featured in *Spinoff*. The Small Business Technology Transfer (STTR) contract is similar to the SBIR, but focuses on commercializing technology developed by universities and federal laboratories through the entrepreneurial efforts of a small business.

Other Organizations

Under a Space Act Agreement, the **Computer Software Management and Information Center** (COSMIC) went from NASA contractor to self-funding business partner in late 1996. COSMIC continues to distribute software for the NASA centers on a cost-recovery basis, and to provide services such as peer review, electronic cataloging, program support, author feedback, and update notification.

The robotics industry is an $8 billion per year international industry, once dominated by the United States. The **National Robotics Engineering Consortium** (NREC) in Pittsburgh, operated by Carnegie-Mellon University, establishes a collaborative government/academia/industry commitment to rapidly move NASA-developed robotics technology out of federal laboratories and into the private sector. A self-sustaining program, the consortium was initiated in 1994 with NASA grants to Carnegie-Mellon which stipulate equal participation by government and industry. NREC aims to re-invigorate the U.S. robotics industry, with projects including the development of an automated agricultural harvesting system and an autonomous mining vehicle.

The **Federal Laboratory Consortium** (FLC) for Technology Transfer promotes and strengthens technology transfer nationwide. More than 600 major federal laboratories and centers and their parent departments and agencies are members, including NASA. **Research Triangle Institute** (RTI) provides a range of technology management services including: technology assessment, valuation and marketing; market analysis; intellectual property audits; commercialization planning; and the development of partnerships. In the last two years, RTI commercialization activities in support of NASA have resulted in 12 new products, 17 cooperative agreements and 24 licenses in product areas that include medical, manufacturing, public safety, sensors and materials.

Managed by the NASA Center for AeroSpace Information (CASI) in Linthicum Heights, Maryland, the **Scientific and Technical Information Program** (STI) provides access to over 3 million aerospace and related citations. STI produces databases such as RECONplus, an electronic version of the NASA Thesaurus, and the NASA GALAXIE system, a database of all holdings in NASA libraries.

There are over a dozen **Commercial Space Centers** (CSCs) in the network, each providing a unique service. The CSC program was established in 1985 to increase private sector investment and interest in commercial space-related activities while stimulating advances in promising areas of research and development. The CSCs include the **Center for Satellite and Hybrid Communications Networks** at the University of Maryland, established in 1991 and specializing in communication networks that utilize both satellite and terrestrial resources.

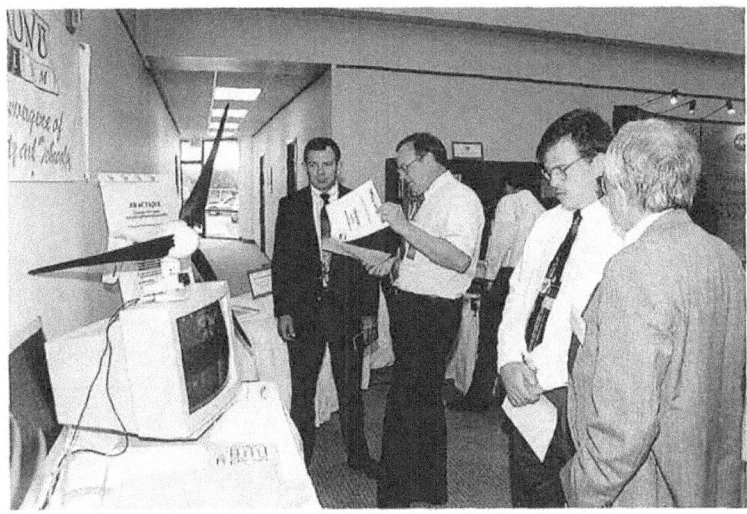

At an open house at the Florida/NASA Business Incubation Center, the president of Coconut Telecomp Inc., an off-site client, speaks to Brevard Community College's Chip Woods about the company's remote sensing device while other attendees look on.

Publications

NASA wind tunnel technique for studying the flow of fluid over a surface by use of multilayered, multicolored coatings helped lead to a breakthrough in one company's medical diagnostic systems.

This application of NASA technology to the commercial sector did not happen accidentally—it was inspired by technology distributed through the monthly *NASA Tech Briefs* publication.

DiaSys Corporation, Waterbury, Connecticut, was experiencing difficulties with its Optical Slide Assemblies (OSAs) used in its automated urinalysis system, R/S 2000. Particles tended to collect and bunch up within cell chambers and obscure the view from the microscopes, rendering the specimens unacceptable. Then Walter Greenfield of DiaSys discovered several applicable articles in *Tech Briefs*, including the wind tunnel technique by Langley Research Center and work performed by the Jet Propulsion Laboratory on hydrodynamic stability.

This led company researchers to aerospace studies on fluid dynamics, such as articles on the characteristics of airflow and how it parallels fluid motion. DiaSys President Todd M. DeMatteo said, "Taking advantage of the information presented, we were able to design the OSA to be aerodynamically—and therefore fluid-dynamically—correct." The company later applied the same technology to its FE-2 workstation, which automates and reduces the cost of microscopic analysis of fecal concentrates.

The National Aeronautics and Space Act requires NASA contractors to provide written reports about inventions, improvements and innovations developed while working for NASA; information from these reports and work done by NASA scientists and engineers are the basis of *NASA Tech Briefs*. About 70 percent of the listings have an accompanying Technical Support Package (TSP), which has detailed information about the particular technology. The briefs are also available on the Internet, with searching capabilities for issues printed over the last two decades. The publication is free and is an awareness and problem-solving tool for U.S. government and industry readers. The articles have ranged in subject from increasing the shelf life of food products to new computer scheduling programs and technology for aircraft drag reduction.

Tech Briefs began in 1962 as a single white paper, which grew into a quarterly government-funded publication printed and distributed by the U.S. Government Printing Office in 1976. The publication evolved

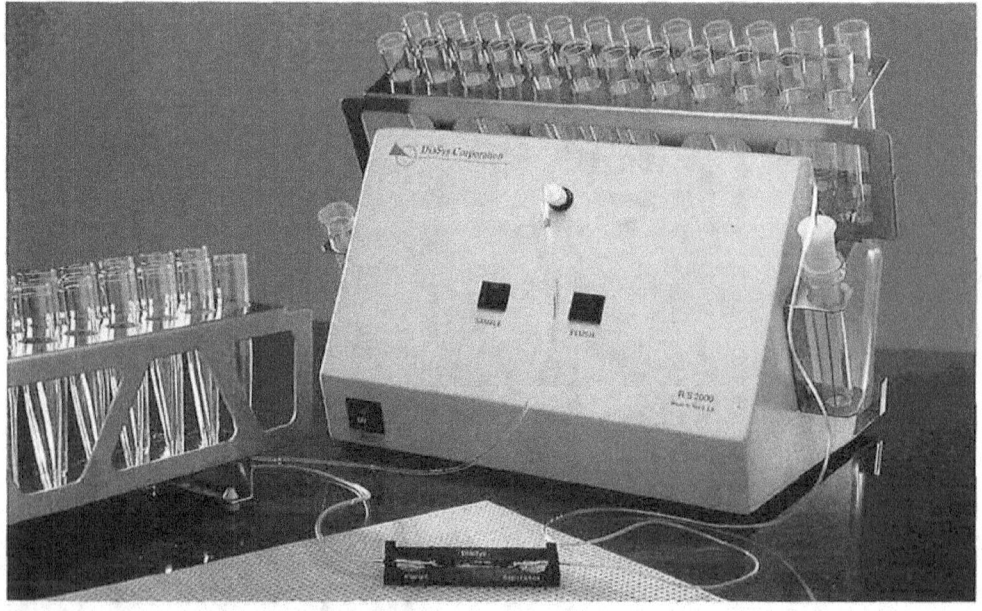

NASA Tech Briefs helped DiaSys Corporation resolve a difficult problem in the development of the company's R/S 2000 instrument for automated urinalysis. The publication is just one way NASA disseminates information on cutting-edge research.

from a black-and-white circular to a full-color magazine, privately-funded and produced commercially by Associated Business Publications Co., Ltd. New York, New York. The venture not only saves NASA all printing and postage costs, but other publication costs as well, totalling approximately $20 million since it began. In January of 1997, the magazine celebrated its 20th year—with over 12,000 individual briefs published, 1.8 million requests for Technical Support Packages generated, and over 207,000 current subscribers.

Complementing *NASA Tech Briefs* is the *Spinoff* publication, which highlights technology transfer successes that result in commercial products or processes. The stories featured include technology transferred through *NASA Tech Briefs*, NASA technical reports, former NASA employees, contractor diversification, technical assistance, patent licenses, and many other mechanisms. Although distributed by the National Technology Transfer Center, the NASA Center for AeroSpace Information (CASI) performs the research and coordination of the publication. Companies with NASA spinoff products or processes may contact Walter Heiland at CASI directly in order to be considered for the publication. CASI also maintains a database of published stories on the Internet which includes abstracts, key centers, origins of the technologies, and manufacturer's addresses. This database and a Web version of the publication can be accessed at **http://www.sti.nasa.gov/tto/spinoff.html**.

Every NASA facility and affiliated organization maintains a site on the World Wide Web, all of which can be accessed through the NASA Commercial Technology Network homepage at **http://nctn.hq.nasa.gov/nctn/**. On-line versions of many other publications including *Aerospace Technology Innovation*, a bimonthly publication of the NASA Office of Aeronautics, and monthly newsletters from the field centers are also linked to the homepage.

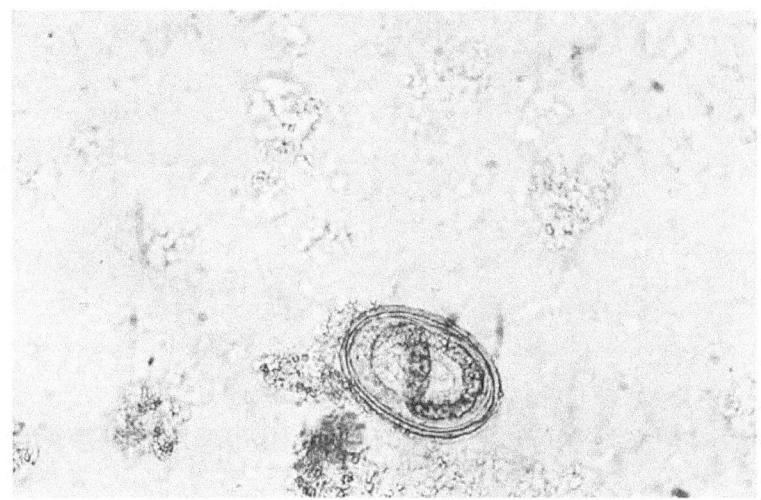

The FE-2 workstation, developed using information from NASA Tech Briefs, automates and reduces the cost of microscopic analysis of fecal concentrates. Shown is a sample containing an ascaris embryo, magnified 100 times, courtesy of Dr. Dwight F. Miller of Saint Mary's Hospital in Waterbury, Connecticut.

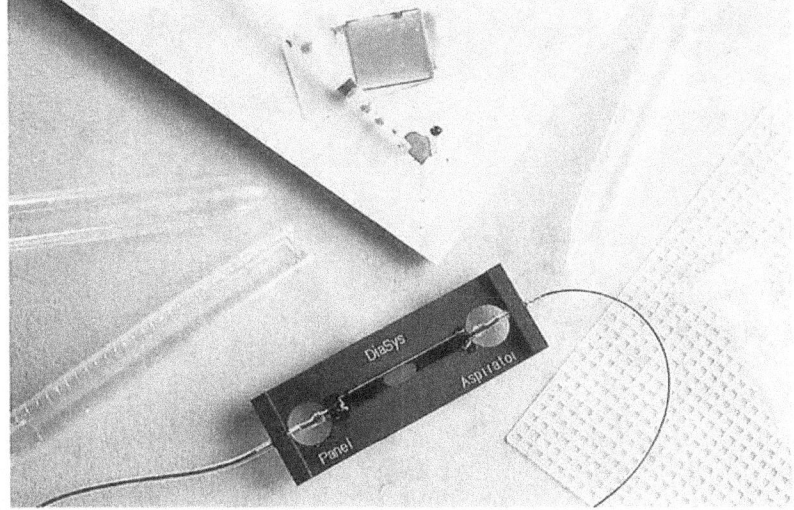

DiaSys Corporation's automated urinalysis system (bottom of photo) eliminates the use of potentially messy and hazardous pipettes, slides and slide cover slips (top of photo). NASA hydrodynamic studies published in NASA Tech Briefs helped the company resolve fluid manipulation problems in the system.

Commercial Benefits—Spinoffs

The idea that if we can go to the Moon we can accomplish other feats long considered impossible has been firmly implanted in people's minds. Confidence that solutions can be found to such urgent problems as an energy shortage, environmental degradations, and strife between nations, has been nourished by this spectacular demonstration in space of man's capabilities.
—*James E. Webb, NASA Administrator 1961-1968*

The NASA technologies that lend themselves to successful commercialization have obviously increased and diversified in the past 25 years. Research in aeronautics; in life, microgravity, and space sciences; in communications; and in space flight and access has brought forth an array of sophisticated technologies ready for transfer to the private sector. Likewise, the resulting products are diverse.

In 1973, this publication featured a new externally-rechargeable pacemaker eliminating costly battery replacement surgery every two years. Today's publication includes a story of interest to other consumers: improvements in suction, vibration, and particle removal by a vacuum cleaner incorporating NASA expertise in a number of disciplines.

The following pages once again highlight spinoff products and services. These wide-ranging benefits are the result of application and development engineers at work in bold and close partnership with NASA and the Agency's Commercial Technology Network.

Muscle Stimulation Technology

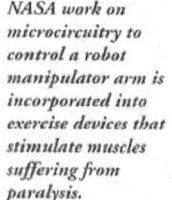

Many a Space Shuttle astronaut has been aided in orbit by the extended reach of the six degree of freedom robot arm, termed by NASA as the Remote Manipulator System or RMS.

Control systems for an RMS simulator have been merged with software and high density hardware to run an electrical stimulation medical device. This fusion of aerospace research and biomedical need is the work of Electrologic of America, Inc., based in Dayton, Ohio.

Electrologic of America (ELA) manufactures several functional electrical stimulation (FES) medical devices. Neuromuscular electrical stimulation, a technique which is commonly referred to as FES, has been used to revitalize purposeful movement to muscles crippled by spinal cord injuries. Former *Superman* star Christopher Reeve, who suffered paralysis after falling from a horse, uses the StimMaster FES Ergometer. Using StimMaster,

paraplegics and quadriplegics can get a full cardiovascular workout equivalent to jogging three miles three times per week.

Under a Goddard Space Flight Center contract, ELA was able to refine the process of densely packing circuitry on personal computer boards. ELA was able to provide significant contributions to Goddard adaptive, closed-loop control systems for the Remote Manipulator System Simulator (RMSS). This required design and fabrication of a new computer-controlled servo system for manipulation of the six-axis, 5,000-lb. mechanical arm which simulates the RMS carried on most Space Shuttle missions.

"With several modifications, we were able to use this type of technology to incorporate it into the software used in the StimMaster FES Ergometer," explains Steven Petrofsky, ELA's Executive Vice-

NASA work on microcircuitry to control a robot manipulator arm is incorporated into exercise devices that stimulate muscles suffering from paralysis.

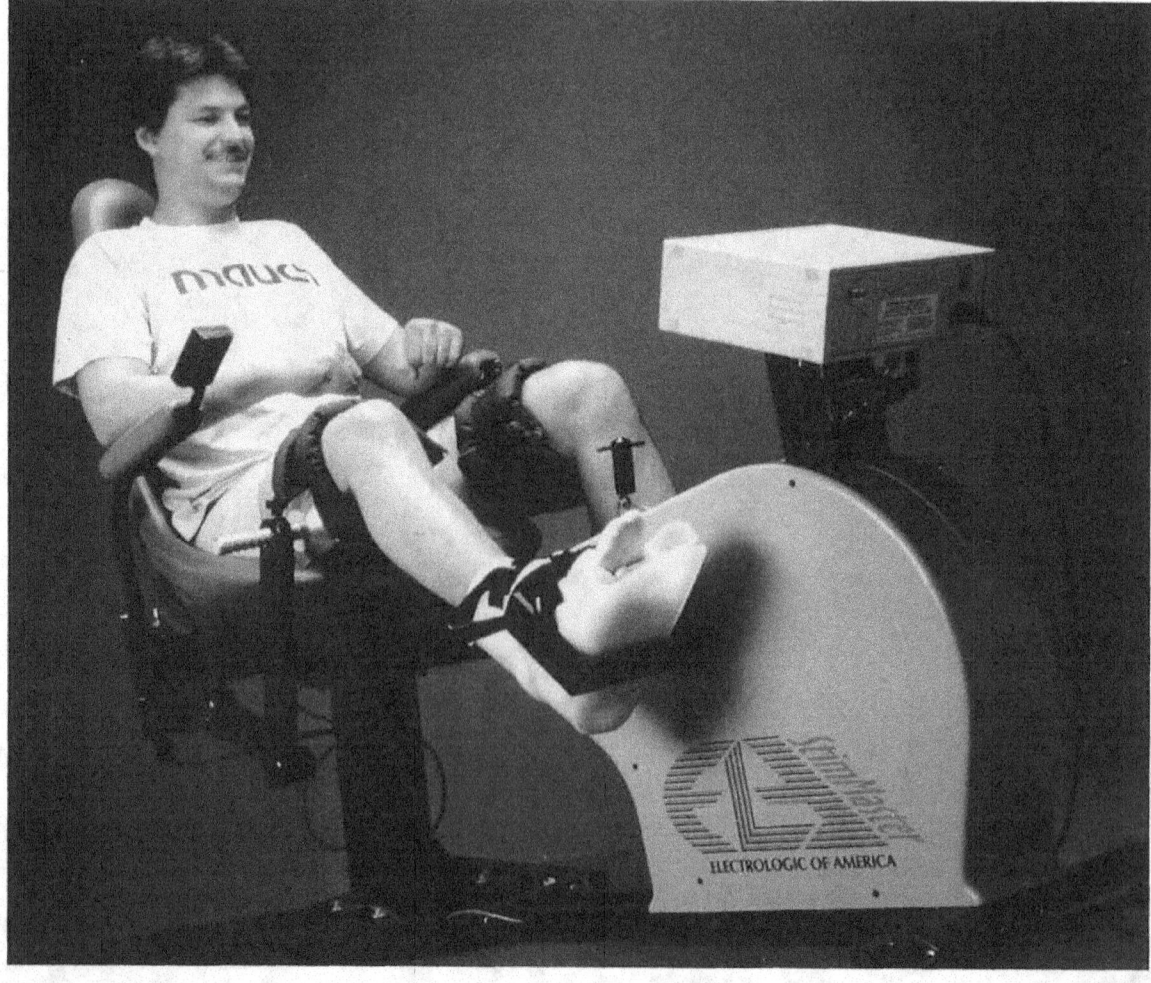

President. He has been the recipient of several NASA awards for outstanding hardware design and robotic control developments, and was instrumental in the software development.

Joe Mica, the NASA RMSS systems engineer and manager, said that Petrofsky's efforts were essential to the success of the RMSS. Ned Conklin of Forth Inc., an ELA subcontractor, implemented mission control software; Robert Lea of Ortech Engineering, Yashvant Jani of Hitachi and Mica together developed the RMSS fuzzy logic control design and published it in the CRC Press industry standard reference book *The Industrial Electronics Handbook*.

The StimMaster is used by persons with paralysis to pedal a recumbent bicycle by stimulating the leg muscles—hamstrings, quadriceps and gluteus maximi—to maintain a consistent rate of 50 revolutions per minute under resistance. Patients steadfastly using the StimMaster Ergometer have experienced diminished secondary symptoms related to paralysis.

"The results of the closed-loop, adaptive control under resistance is the reversal of atrophy, improved circulation and the relaxation of muscle spasms," Petrofsky adds.

The StimMaster incorporates sensors, located within the ergometer, that provide continuous feedback to a computer. This computer controls the rate of pedaling through muscle stimulation, thereby achieving a rhythmical pedaling motion. Because the units are designed for home as well as clinical use, a person suffering from spinal cord injury can carry out a therapy program in the privacy of their own residence.

StimMaster's advanced computer continually monitors the patient's progress every 1/40th of a second and adjusts the settings to meet the patient's needs.

ELA's work for NASA on computer circuitry has also been applied to the VST-100. This portable, electrical stimulation equipment was developed by Petrofsky exclusively for V-Care Health Systems, Inc., based in Washougal, Washington.

The state-of-the-art VST-100 can increase bloodflow to afflicted areas, rejuvenate muscles and improve recovery time of an injured person—all through electrical stimulation. When used by a person distressed by carpal tunnel syndrome, as example, the VST-100 administers electrical pulses that increase circulation in the wrist, which opens up nerve pathways. Using the muscle stimulator technology, a person can return to the job more quickly, work productivity is increased, and health-care costs are decreased.

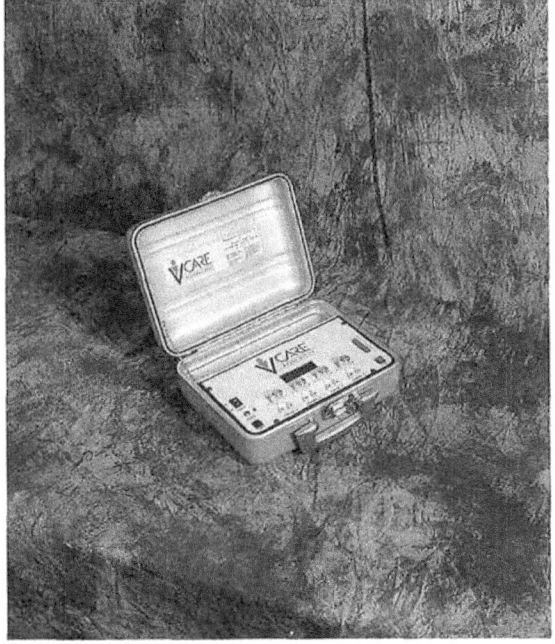

VST-100 power muscle stimulator can improve the bloodflow to injured areas. The equipment uses electronic microcircuitry spurred by NASA research on robot remote manipulator systems.

Gold Coating

Reflect on this. Many satellites carry gold-coated mylar sheets to protect them from solar heat.

A thin layer of gold on an astronaut's helmet visor fends off dangerous effects of solar radiation. Satellite microelectronics that instantaneously relay data around the globe depend on gold components to ensure reliable, corrosion-resistant and static-free performance.

The growing use of gold in advanced technologies such as microelectronics, telecommunications, optics, aviation and space has increasingly made gold a vital strategic resource in U.S. technological and economic competitiveness.

In 1996, the Mars Global Surveyor blasted off toting a gold-plated telescope mirror, part of a laser device that is to chart the topography of the entire Martian surface over a two year period.

Epner Technology Inc. of Brooklyn, New York rose to the challenge of a NASA Goddard Space Flight Center requirement for the ultimate in electroplated reflectivity needed for the Mars Global Surveyor's Mars Orbiter Laser Altimeter (MOLA). The MOLA mirror, an unusually large one-half meter in diameter, was ground by OCA Applied Optics in Garden Grove, California. Made of beryllium, the MOLA mirror was coated by Epner Technology's Laser Gold® process, specially improved for the project. The resultant mirror coating proved exquisite.

Laser Gold is a proprietary process for electrochemically-deposited gold. In the near infrared wavelengths, Laser Gold's reflectance is an astounding 99.4 percent.

"We are not paranoid about our processes since the real secret in this business is in the controlling of those processes," says Epner Technology owner and CEO, David Epner. "Our strength comes from a lot of years of breaking the back of some really off-beat plating challenges," he says.

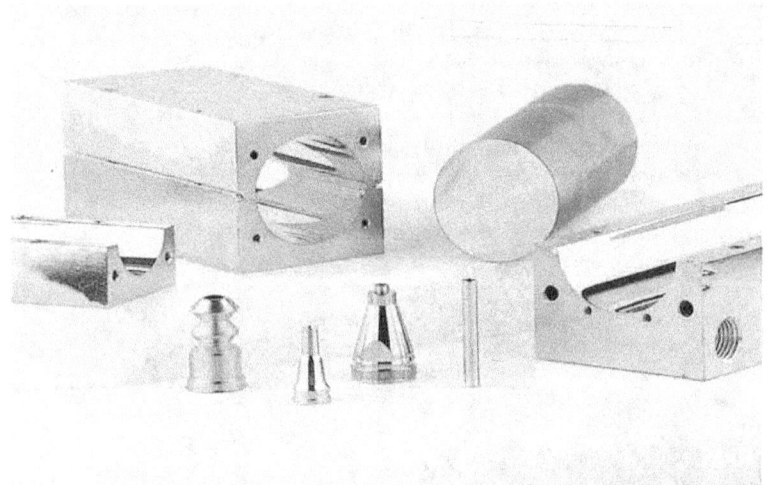

Epner Technology Inc.'s Laser Gold process is used to coat products, including lasing cavities, used in various surgical instruments. Epner improved its coating process based on modifications it made for a mirror on the Mars Global Surveyor.

Using the Laser Gold coating, Braun's temperature sensor provides rapid, accurate results.

The gold coating on the MOLA mirror is a case in point. To give the telescope mirror the needed sensitivity, the Laser Gold coating was essential to mapping operations around Mars. Gold's extremely high-quality reflectivity is critical in the capturing of laser infrared radiation that is bounced back from Mars to the mirror's surface.

"What makes Laser Gold incomparable for the most sophisticated optical and laser applications like the Mars laser altimeter, where the margin for error is absolutely nil, is its extraordinary reflectivity, in combination with its cleanability and virtually perfect reliability and corrosion resistance," Epner says.

Thanks to the NASA push, adds Epner, improved Laser Gold-coated reflectors have found use in an epitaxial reactor built for a large semiconductor manufacturer. The reactor heats the silicon wafers inside a quartz bell jar with infrared energy from some 100 six-kilowatt quartz-halogen lamps. Behind these lamps are an array of Laser Gold-coated water-cooled aluminum extruded reflectors.

The improved reflectivity of these reflectors has dramatically increased lamp life, says Epner, due to the lower power requirements that the reflectors permit. The huge power consumption of these machines was also reduced with this increased efficiency. These reflectors will be sold worldwide to firms like Motorola, Epner says.

Once again, due to NASA's demanding quality needs, Epner's Laser Gold coating has also found use as a waveguide in Braun-Thermoscan's tympanic thermometer.

Epner Technology customers are the foremost fabricators in aerospace, defense, microwave and electronics, optics and semiconductors.

® Laser Gold is a registered trademark of Epner Technologies, Inc.

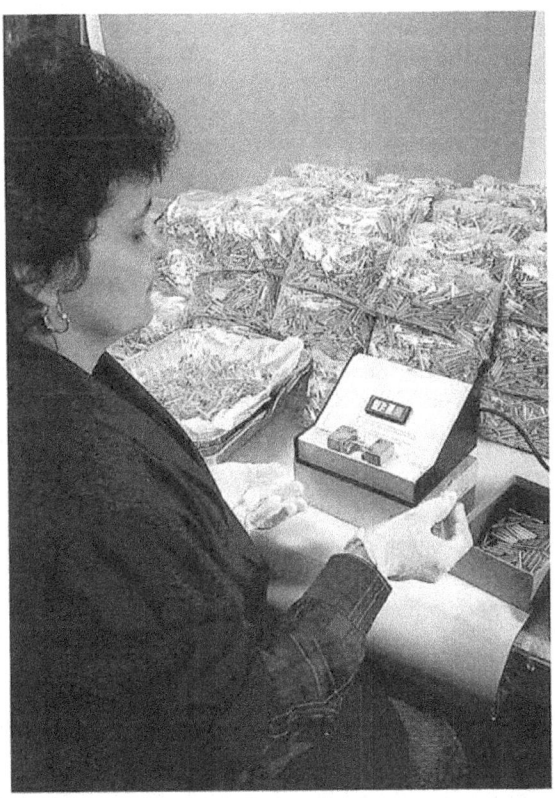

Each Laser Gold-coated tube used for the tympanic thermometer is inspected for reflectivity and dimensions by hand.

Automated Analysis Workstation

Automated microscopic analysis of fecal concentrates can be performed at reduced cost, higher accuracy and with enhanced worker safety by a new counter top clinical workstation. DiaSys Corporation of Waterbury, Connecticut has developed the FE-2, an easy-to-use workstation for rapid location and identification of parasites, cysts and ova in human feces.

Several years ago, information appearing in the pages of the *NASA Tech Briefs* publication assisted DiaSys's manufacturing of the R/S 2000, the company's first product. There is now a family of "R/S" workstations, all of which speed-up, standardize, automate and make safer the laboratory analysis of urine sediment. The initial "R/S" workstation was made possible in part by coverage of work at the Jet Propulsion Laboratory and Langley Research Center. This stimulated a fast-paced roll-out of additional products from DiaSys.

The success of the R/S 2000 was followed by the R/S 1000, another product that standardizes automated urine sediment analysis for small hospital labs, physician group practices, "stat" labs, and outpatient clinics.

In September 1995, DiaSys introduced its third product, the R/S 2003. This larger, more advanced system standardizes and drives the cost out of routine urine sediment analysis conducted by larger labs and lab networks. The DiaSys workstations have been named the preferred practice by SmithKline Beecham Laboratories, Kaiser Permanente of Southern California, and many other hospitals and private lab groups.

Recently, a fourth workstation product was released to market—the FE-2. This latest technology is a workstation that automates the process of making and manipulating wet-mount preparations of fecal concentrates. A decrease in time needed to read the sample is achieved, permitting technologists to rapidly spot parasites, ovum (eggs) and cysts, sometimes carried in the lower intestinal tract of humans and animals.

The FE-2 workstation corrects the procedural shortcomings of current methodologies in several significant ways. First of all, no special training is required to operate the FE-2. To operate, the technologist inserts the instrument's automatic dual aspirator

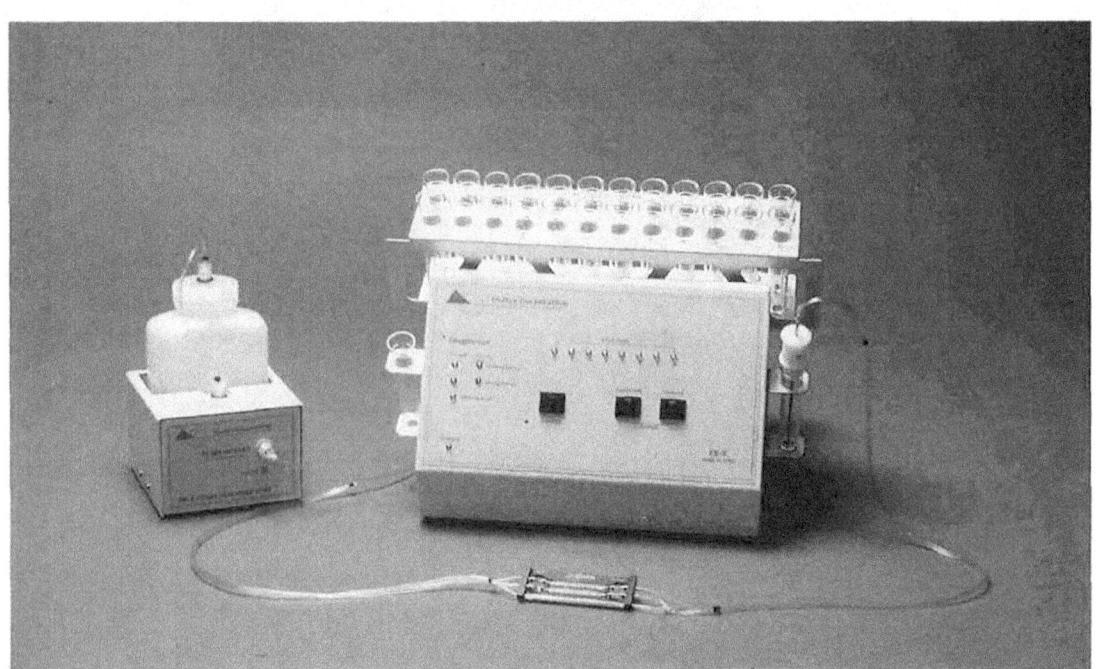

DiaSys Corporation's FE-2 workstation eliminates the need and cost of disposable pipettes, microscope slides, and cover slips for faster and safer analysis of fecal concentrates.

into a prepared fecal concentrate and presses the "sample" button on the workstation's control console. Within 5 seconds, a consistent, measured amount of concentrate is automatically transferred to the workstation's optical slide assembly (OSA) on the stage of the microscope.

During this process, one chamber of the OSA is inoculated with a small quantity of the fecal concentrate. The other chamber is inoculated with an equal mix of fecal concentrate and iodine (for staining) or isotonic saline (for dilution) as the lab prefers. Observations of the stained (or diluted) and unstained fecal concentrate are then ready to begin. Purging the system takes about 7 seconds, after which time the FE-2 is ready to make the next concentrate slide.

Among its virtues, the FE-2 increases the level of safety and precision by automating the aspiration, resuspension, staining or diluting, transfer, presentation, and disposal of fecal concentrates. Eliminated is the need and cost of disposable pipettes, microscope slides, and cover slips. Moreover, because the fecal concentrate

is contained in and manipulated through a sealed system, the technologist carrying out the work is not exposed to prolonged inhalation of chemicals such as formalin which are conventionally used to process raw fecal material.

Employing the FE-2 is non-invasive, can be performed on an out-patient basis, and quickly provides confirmatory results, says DiaSys Corporation president, Todd DeMatteo. "To the best of our knowledge, there is no competitive product or system in the market today," DeMatteo says.

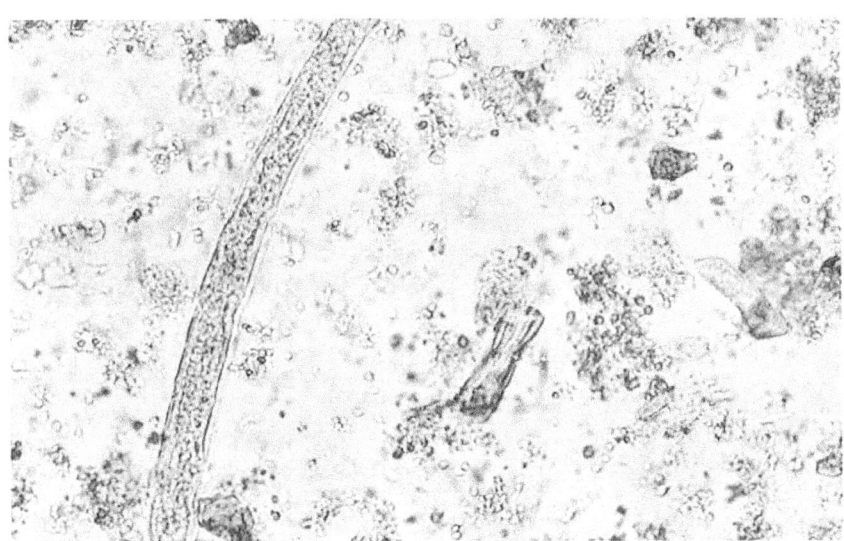

(Specimen courtesy of Dr. Dwight F. Miller at Saint Mary's Hospital in Waterbury, Connecticut.) Parasites in a fecal sample (at a magnification of 100 power) can be detected by the FE-2.

Innovations in Aircraft Design

When the Boeing 777 first took wing, it carried with it basic and applied research, technology, and aerodynamic knowledge honed at several NASA field centers.

The precedent-setting 777 was built to handle medium- to long-range passenger flights and is the largest twin-engine jet to be manufactured today. First passenger-carrying flights began in May 1995. According to Boeing Company estimates, the 777 fleet has captured three-quarters of new orders for airplanes in its class since the program was launched.

In May 1996, the first Boeing 777 stopped by Langley Research Center as a salute to NASA's involvement in its creation. Several Langley innovations were instrumental in the development of the aircraft, such as:

- fundamental mathematical procedures for computer-generated airflow images which allowed advanced computer-based aerodynamic analysis;
- wind tunnel tests, confirming the structural integrity of 777 wing-airframe integration in Langley's Transonic Aerodynamics Tunnel. Use of the facility was reimbursed to NASA by Boeing;
- knowledge of how to reduce engine and other noise for passengers and terminal area residents;
- radial tires that are used on the aircraft underwent strength and durability testing at Langley's Aircraft Landing Dynamics Facility; and
- increased use of lightweight aerospace composite structures for increased fuel efficiency and range. The 777's floor beams, flaps and tail make use of lightweight composites.

At Marshall Space Flight Center, results from tests aimed at improving the performance of NASA's Space Shuttle engines led to improvements in the Boeing 777's new, more efficient jet engines. Working with Pratt & Whitney, the U.S. aircraft and rocket engine provider, NASA engineers conducted evaluations of wake patterns flowing through the plane's turbine engine airfoils. Data taken proved useful in obtaining better turbine efficiency, as well as realizing substantial fuel savings.

Boeing 777 inlet, hinge and strut blankets were quilted with either stainless steel or ceramic thread. These blankets protected areas of the plane from high temperatures and fire. Fostered by Ames Research Center, the Boeing 777 blankets have a lineage to Advanced Flexible Reusable Surface Insulation (AFRSI) used on certain areas of the Space Shuttle.

Several other areas benefitted from NASA and Boeing collaboration. Langley had contracted with Boeing, for example, to design and validate a digital flight control system for fly-by-wire and fly-by-light/power-by-wire applications. In developing the digital fly-by-wire system, researchers utilized the Apollo guidance, navigation and control hardware as the primary digital system. Fly-by-wire systems for control of wing and tail surfaces replace bulkier and heavier pulley and cable systems on the 777.

The Boeing 777's modern glass cockpit is a system that uses computer technology to integrate information and display it on monitors in easy-to-use format. Research was undertaken on the challenge of maintaining a pilot's situational awareness during flight operations. The evolution of the highly automated glass

Building the Boeing 777 brought about the use of NASA innovations, from lightweight composite materials to the modern glass cockpit and aircraft control systems.

cockpit, particularly in commercial aircraft, has roots at Langley Research Center.

Taken as a whole, NASA contributions in fundamental research and technologies proved meaningful to 777 development. Together, industry and government skills and abilities melded, jointly contributing to the airplane's operating efficiency, passenger service, environmental compatibility and safety.

Attitude Control

Spacecraft stability and control is a must so that sensors, solar panels, antennas and other hardware are properly oriented to perform their functions. In order to satisfy a demand for low cost attitude control systems for small spacecraft, Goddard Space Flight Center awarded Small Business Innovation Research (SBIR) contracts to ITHACO, Inc. of Ithaca, New York.

By taking on that challenge, ITHACO designed and built versatile but inexpensive scanning horizon sensors and reaction wheel hardware. By transferring that expertise into commercial products, the company has become a prominent manufacturer of attitude sensing and control components in the United States and for the international community.

A pioneer in the field of satellite control since 1962, ITHACO has produced such items as Earth sensors, reaction/momentum wheels, magnetometers and magnetic torquers. Over 100 U.S. satellites have carried equipment crafted by the firm, including spacecraft flying under Japanese, Canadian, French, German,

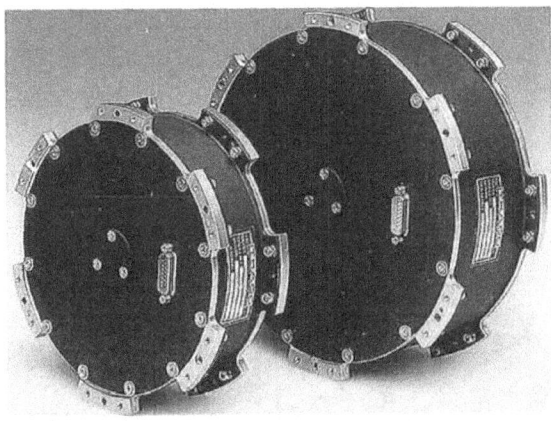

ITHACO's low-cost but highly reliable reaction wheel designs keep spacecraft correctly oriented as they spin through space.

Spanish, Swedish, and Argentinean flags. That track record fits well with ITHACO mottos: "Winning Attitudes In Space" and "Attitude Is Everything."

NASA SBIR-sponsored work resulted in the ITHACO's T-Wheel, built specifically for attitude control of small and medium-sized spacecraft. The T-Wheel consists of a precision balanced aluminum flywheel suspended on stainless steel ball bearings and driven by an ironless armature brushless DC motor. One exceptional feature of the T-Wheel is that it can be used as an independent momentum/reaction wheel or augmented with horizon sensor optics for both attitude determination and attitude control.

Another ITHACO product is the T-SCANWHEEL®. T-SCANWHEEL's mixture of attitude determination and control capacity reduces overall system cost, minimizes mass and power, and enhances reliability.

Additionally, ITHACO has produced the Type E Wheel. This highly reliable hardware for reaction torque and angular momentum storage for attitude control is built for use on medium to large spacecraft. An integral motor in the Type E Wheel accelerates or decelerates a flywheel as a way to control momentum exchange with the spacecraft. This is valuable for a variety of attitude control schemes.

Located in upstate New York, ITHACO recently opened its 40,000 square foot facility, complete with a class 10,000 clean room. The facility was constructed to hold engineering, manufacturing, test, and corresponding support functions within a single building.

The first T-Wheel contract was signed in 1989. Through July 1996, ITHACO had delivered or was under contract for 95 T-Wheel, T-SCANWHEEL and Type E Wheel units. Sales projections for the hardware are indeed encouraging given the burgeoning small satellite industry. The potential market for the ITHACO wheels is directly proportional to that for small and medium size satellites. Bullish on its past performance and products, ITHACO expects to capture at least 25 percent of this market, with the potentiality of increasing their slice of space business to 50 percent.

ITHACO maintains an active internal research and development effort. New and innovative attitude determination and control hardware includes a lunar horizon crossing indicator and an Earth sensor for geosynchronous applications. Company research dollars are being spent on products that sport increased torque, extended life, and precision balance...definitely a company with an attitude toward the future.

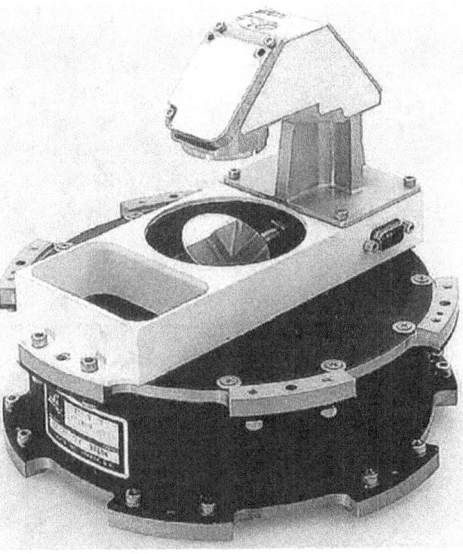

ITHACO's T-SCANWHEEL is a momentum/reaction wheel comprising a high accuracy Conical Earth Sensor to maintain a satellite's precise attitude.

® T-SCANWHEEL is a registered trademark of ITHACO, Inc.

Reaction/Momentum Wheel

Being a big wheel in spacecraft lingo may be detrimental if you are trying to stay competitive in the explosive world of small satellite building.

CTA Space Systems, Inc. of McLean, Virginia has been licensed to sell commercially a reaction/momentum wheel originally developed for NASA's scientific satellites. The wheel is unique in its small size, extremely low residual imbalance, and large, highly controllable torque. These are the same features required by a growing number of small commercial satellites.

In licensing from NASA, CTA was able to virtually eliminate the technical risk and minimize the financial investment that would normally be required to introduce a new spacecraft component.

Reaction/momentum wheels are flywheels used to provide attitude control authority and stability on spacecraft. By adding or removing energy from the flywheel, torque is applied to a single axis of the spacecraft, causing it to react by rotating. By maintaining flywheel rotation, called momentum, a single axis of the spacecraft is stabilized. Several reaction/momentum wheels can be used to provide full three-axis attitude control and stability.

Inside CTA Space System's High Torque Reaction/Momentum Wheel is an innovative flywheel/bearing arrangement that allows the entire rotating system to be balanced after it is assembled.

NASA originally identified a need for the wheel in its Small Explorer (SMEX) program — an initiative to develop highly focused and relatively inexpensive scientific spacecraft. One such spacecraft is the Submillimeter Wave Astronomy Satellite (SWAS). Keeping the SWAS instrument pointed at celestial objects to a high degree of precision (arc seconds) requires extremely low jitter of the spacecraft. SWAS also required a reaction/momentum wheel with a torque greater than any comparably sized commercially available wheel.

Those requirements prompted Goddard Space Flight Center to design and build a Small, High Torque Reaction/Momentum Wheel. This wheel is unique in that it features a large, highly controllable torque in a small package with low power and an extremely low

residual imbalance. To minimize imbalance, an innovative flywheel/bearing arrangement is used that allows the entire rotating system to be balanced after it is assembled.

The development has proven so successful, eight more wheels have been fabricated by NASA for other SMEX space science missions.

A market assessment by Research Triangle Institute (RTI) for Goddard Space Flight Center was completed, showing that the flywheel technology had commercial potential. NASA applied for a patent and pursued commercialization of the technology with RTI assistance.

CTA had prior experience with the technology while under government contract. Recognizing that the Small, High Torque Reaction/Momentum Wheel developed by NASA had unique specifications, CTA applied for a patent license, with an eye toward commercial sales. The license was granted in the fall of 1996, and CTA Space Systems, Inc. currently uses the technology in its complete spacecraft fabrication services.

CTA Space Systems has successfully built over 10 of the reaction/momentum wheels for commercial, scientific, and military customers. Engineers from Goddard Space Flight Center are continuing to work with CTA to transfer and adapt the technology for wider commercial use and application.

As a leading manufacturer of lightweight satellites, CTA Space Systems must develop spacecraft and integrate payloads in short time frames to assure standards for affordability. The reaction/momentum wheel technology is sure to help maintain that thrust.

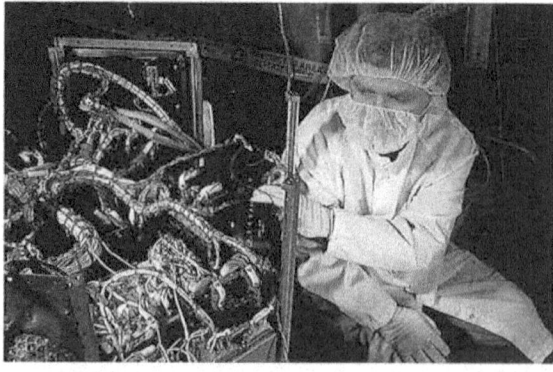

Eli Ahronovich, Principal Mechanical Engineer at CTA Space Systems, connects a momentum wheel to a commercial space satellite. Several reaction/momentum wheels can be used to provide full three-axis attitude control and stability.

Wing Flutter Control

O rbital Research Inc. of Cleveland, Ohio has developed a software program that "thinks" change. By way of Small Business Innovation Research (SBIR) contracts from Langley Research Center, an Orbital Research Intelligent Control Algorithm™ (ORICA™, rhymes with "eureka") has been developed. It is the first practical hardware-independent adaptive and predictive control structure, specifically suited for optimal control of complex, time-varying systems.

Orbital Research has applied ORICA technology to the problem of controlling aircraft wing flutter, movement that is akin to a shimmy in an automobile. The effectiveness of the control software has been validated during wind tunnel testing at Langley. In related work, Orbital Research and NASA's Lewis Research Center have determined that the company's controller could be coupled with Lewis' high speed valve technology to improve aircraft performance.

The company's technology, coupled with NASA expertise, has the possibility of making jet travel safer, more cost effective by extending distance range, and lowering overall aircraft operating costs.

There are three main components to ORICA. At its heart is an Identifier Module which estimates a mathematical model of the system that is being controlled. The model, which is re-identified at every time step, relates the effect of each actuator input to the output of the system. The mathematical model estimated by the Identifier Module is then used by the Predictor Module to predict the future condition of the system to be controlled. Finally, the Controller Module uses the estimated future response of the system developed by the Predictor to calculate the actuator's optimal position to meet the desired response.

To a pilot, large changes in control before action takes place, could mean the difference between life and death. The ORICA controller doesn't wait to take action. Rather, the intelligent software program anticipates and makes adjustments beforehand. As shown in Langley wind tunnel testing, the ORICA can control a small flap on a wing, nullifying flutter by regulating the flap's angle as many as 500 times per second.

Because of its computational efficiency, the superior performance of ORICA was achieved by employing a personal computer platform (Intel® Pentium®) as opposed to more expensive hardware commonly tasked for these high speed flutter suppression tests. ORICA's special algorithms require only one percent of the computations of other more expensive and complex academic controllers.

One of ORICA's first applications was in suppression of flutter of a wing subjected to speeds up to Mach 0.95 at Langley's Transonic Dynamics Tunnel. ORICA successfully reduced vibrations on the wing by an order of magnitude over a standard fixed gain controller.

The commercially available ORICA software is not strictly for use on aircraft. Future application areas for ORICA include control of robots, power trains, systems with arrays of sensors, or regulating chemical plants or electrical power plant control.

™ Orbital Research Intelligent Control Algorithm and ORICA are trademarks of Orbital Research Inc.
® Intel and Pentium are registered trademarks of Intel Corporation.

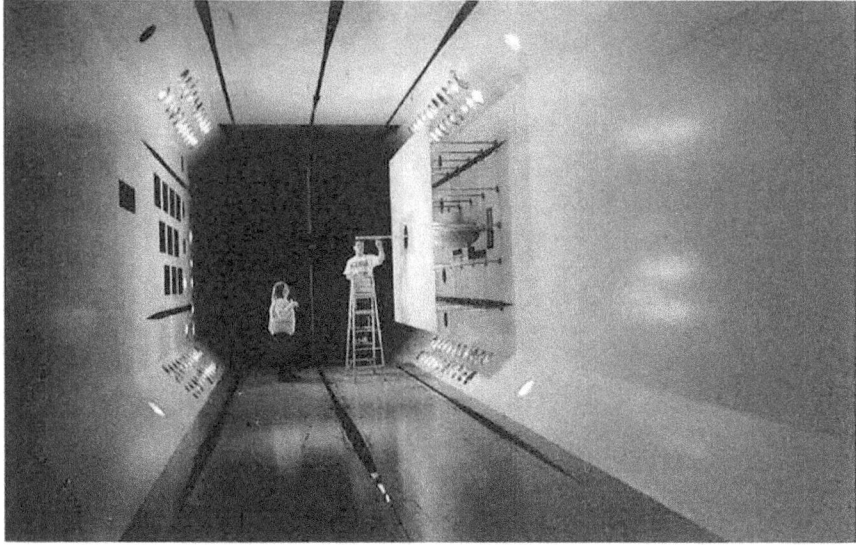

Wind tunnel testing of ORICA intelligent control software has shown it to be highly responsive in damping out wing flutter. Application to commercial aircraft could result in safer and more cost-effective jet travel.

Reusable Surface Insulation

Searing heat engulfs a Space Shuttle orbiter as it plunges through the atmosphere en route to a touchdown on Earth. Protecting select areas of the space plane during its fiery dive is Advanced Flexible Reusable Surface Insulation (AFRSI).

AFRSI was developed by Ames Research Center then integrated into the Space Shuttle by Rockwell International. Production of the AFRSI was transferred in 1974 to Hi-Temp Insulation Inc. of Camarillo, California.

For over 22 years, Hi-Temp has provided insulation blankets for the external leeward surfaces of the Space Shuttle. Hi-Temp created many new technologies to meet the requirements of the Space Shuttle program. A combination of low weight and high thermal efficiency demanded an examination of new materials and fabrication techniques. Material must resist temperatures of some 2,000 degrees Fahrenheit for 15 minutes with no burn through.

On low-temperature areas of a Space Shuttle, the company made molded fiberglass insulation covered with polyester film. Hydraulic lines and system components are wrapped with this insulation.

A Shuttle orbiter's payload bay is protected from heat by using Hi-Temp's two-blanket insulation system. Lastly, the Space Shuttle's main engine nozzles feature ceramic insulation. This insulation is constructed to withstand temperatures to 2,600 degrees Fahrenheit and extreme vibration.

Improvements to the insulation material by Hi-Temp with the assistance of NASA gives it the ability to withstand heating and cooling cycles; rapid and fluctuat-

Insulation technology based on protecting the Space Shuttle during reentry has been adopted by the aircraft industry and other commercial interests. A quilting machine is used to fabricate thermal blankets for the Space Shuttle.

ing temperature changes; continuous vibration and gravitational stress; and contact with aircraft engine contaminants.

The outstanding virtues of Hi-Temp's range of insulation products have enabled the firm to use this innovation in many other ways. These include: insulation blanket to cover aircraft parts, such as engine exhaust ducts; fire barrier material to protect aircraft engine cowlings; molded fiberglass insulation blankets for acoustical protection; and in aircraft rescue fire fighter suits.

Hi-Temp's Space Shuttle work has nurtured a unique proficiency, with many new production techniques now available that can be geared to both aerospace and commercial markets.

Working closely with material manufacturers, Hi-Temp has fabricated blankets using carbon, graphite, ceramic, and nicalon. To protect and seal sewn insulation blankets, Hi-Temp uses a wide variety of facing materials including coated fabrics, mylars, and films. In addition, using stainless steel thread, foils up to .002 thick can be sewn to insulation blankets for liquid proof protection at high operating temperatures. This skill has been applied to such critical programs as the Trident missile, Delta launch vehicle, as well as the Space Shuttle.

Equipment to fulfill tasks at Hi-Temp include many special machines such as rigidizing rollers for strengthening metal foils; a special heated platen press for bonding structural components; heat-sealing machines; and welders for stainless steel and inconel.

A Fire Protection Division of Hi-Temp Insulation has been established, offering the first suit designed exclusively by and for aircraft rescue fire fighters. Offering great mobility, comfort and flexibility, the fire fighting suit uses proprietary insulating materials that enable the suit to weigh far less than other heat-protecting apparel. From the inside out, the Hi-Temp fire fighting suit retards heat, scalding steam, flammable fumes, and molten liquids.

Hi-Temp is a supplier to the Los Angeles City Fire Department as well as other major U.S. civil and military fire departments.

At its start in the mid-1970s, Hi-Temp relied heavily on government projects, becoming 90 percent government dependent. It is now estimated that this growing concern has become 85 percent commercial dependent, with about 15 percent of its work done under government contract.

Hi-Temp's insulation is used on the Boeing 777. It is assembled on a check fixture and includes a quilted insulation core.

Aviation Design Software

There is an urgent need to help revive an ailing domestic general aviation industry in the United States. That imperative is receiving increased attention by NASA.

To this end, by way of a Langley Research Center Phase I Small Business Innovation Research (SBIR) contract, DARcorporation of Lawrence, Kansas has made the art of designing a general aviation aircraft far easier and less expensive.

Today's elite corps of aircraft manufacturers utilize powerful computers capable of running expensive Computer Aided Design (CAD) software. But the majority of small General Aviation manufacturers cannot afford these investments. Rather, small firms design on paper or with computers utilizing self-generated programs on spreadsheets.

DARcorporation has developed a General Aviation CAD package. This affordable, user-friendly preliminary design system for General Aviation aircraft runs on the popular 486 IBM-compatible personal computers. The system gives the design engineer the tools to briskly evolve an aircraft configuration from weight sizing and wing loading to stability and control surface requirements, among a host of parameters.

A Phase II SBIR contract was awarded by NASA to the company, resulting in a commercial product that the firm is now marketing. Individuals who are taking the home-built approach, small manufacturers of General Aviation airplanes, as well as students and others interested in the design and analysis of aircraft are possible users of the package.

DARcorporation's General Aviation Computer Aided Design (G.A.-CAD)

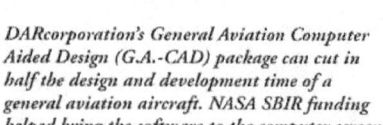

DARcorporation's General Aviation Computer Aided Design (G.A.-CAD) package can cut in half the design and development time of a general aviation aircraft. NASA SBIR funding helped bring the software to the computer screen.

package can reduce design and development time by 50 percent, replacing tedious hand calculations. Typically, design cycle times of two to four years are required to fully conceive a General Aviation aircraft. G.A.-CAD can reduce this time period by half, notes William Anemaat, Vice President of DARcorporation.

The G.A.-CAD enables a user to analyze an existing design, make small changes to one part of the aircraft, and then rapidly determine the effect of the modification to all other aspects of the plane design. All applicable performance and flying quality regulations are embedded in the program. This gives the designer an immediate judgment as to whether or not the design meets air worthiness regulations. Such a feature, therefore, benefits those manufacturers with little experience in obtaining General Aviation aircraft certification.

Engineering Systems, Inc. (ESI) of Colorado Springs, Colorado, is using the DARcorporation's software for aircraft analysis. A small consulting firm in simulator development and a broad spectrum of other engineering disciplines, ESI used the software in the development of an aerodynamic model of the Airbus A320 jet transport for a training device. "We are pleased with the program and have found that the initial cost was quickly repaid in reduced labor costs," said ESI's Director of Aeronautical Engineering, Dr. David L. Kohlman.

Modeled after DARcorporation's work on Advanced Aircraft Analysis (AAA) software, the G.A.-CAD package features ten discrete modules: weight, performance, geometry, aerodynamics, thrust/power, stability and control, dynamics, loads, structures, and cost. These features improve product design. More alternative designs can be evaluated in the same time span, which can lead to improved quality.

By using the cost module, various expenses in airplane design programs can be computed. Research and development, manufacturing, prototype, and direct and indirect operating costs can be estimated using this module.

G.A.-CAD incorporates a user-interface based on the Microsoft Windows® operating system, making the software package highly user-friendly.

Having the software package at various General Aviation manufacturing sites, DARcorporation expects to see significant improvement of competitiveness, and should bolster the chances of success for newly established companies.

® Windows is a registered trademark of Microsoft Inc.

Satellite Antenna Systems

New ways to bring space down to Earth is a success story from KVH Industries, Inc. of Middletown, Rhode Island. A partnership between KVH Industries and the Jet Propulsion Laboratory (JPL) is hastening the day of mobile reception of television via satellite on moving vehicles, such as buses and trains. In addition, successful development of a mobile satellite communications antenna may enable persons on the move to have additional access to the information superhighway.

KVH initially contacted the Rhode Island Technology Transfer Center who, in turn, linked KVH with the NASA laboratory. Discussions between JPL and KVH led to an exclusive licensing agreement that permitted the company to convert space agency hi-tech communications gear into consumer products.

"NASA's satellite technology fits in perfectly with our own satellite technology," said KVH president Martin Kits van Heyningen. "Over the years, we have developed a line of satellite-aiming antennas for marine industry. This latest development with NASA will help extend our product line."

KVH is widely known for its electronic digital compass technology. The company manufactures products for recreational and commercial marine markets, and is a major supplier to the U.S. military of digital compass systems for ships and armored vehicles.

NASA had developed an experimental, microprocessor-controlled satellite antenna for motorists to send and receive phone, fax and other telecommunications as part of the Advanced Communications Technology Satellite (ACTS) program. Tests proved that sensors within the unit kept the antenna continually pointed at the ACTS positioned in space. JPL had developed the tracking antenna system for use on ACTS which can handle digital television signals in the Ka-band frequency from a mobile or parked vehicle.

Through the Technology Affiliates Program at JPL, the ACTS antenna system was transferred from experimental testing status to commercial development at KVH.

The ACTS design enables mobile satellite antennas to remain pointed at the satellite, regardless of the motion or vibration of the vehicle on which it is mounted. This design permits a satellite antenna to smoothly lock onto the desired satellite signal, without encountering aiming fluctuations. Changes in aiming are a common drawback in other land-mobile satellite communication antennas.

Successful tests had shown the ACTS antenna ideal for many applications, including:
- one-way transmission of patient data from field paramedics to the base hospital;
- remote medical imaging including x-rays from a moving vehicle to a fixed station;
- satellite communications for on-the-move military applications; and
- transmission of video communications from a satellite news gathering vehicle to a news bureau.

KVH's first product based on the ACTS design is a land-mobile satellite antenna system that will enable direct broadcast satellite (DBS) television aboard moving trucks, recreational vehicles, trains and buses. KVH's DBS system will provide the link for users on the move to watch multi-channel, high resolution satellite television.

The company foresees other products incorporating the ACTS antenna work, for use in broadcasting, emergency medical and military vehicles. KVH sees a market of several thousand units sold annually and further opening new vistas for telecommunications. Marketing the mobile satellite antenna design is expected to enhance the company's position in the worldwide, mobile satellite communications market.

KVH Industries, Inc.'s satellite antenna improves communication on marine or mobile land transportation, such as trucks, buses, and trains.

Aircraft Flutter Testing

A slogan for Dynamic Engineering Inc. (DEI) is "mobilizing minds and materials in research and development for land, sea, air, and space." That blend of talent and technology came together in 1972 to establish DEI of Newport News, Virginia. The core of early work by the company was building aircraft models for wind tunnel testing at Langley Research Center.

During his 34-year employment at Langley, Wilmer Reed gained international recognition for his innovative research, contributions and patented ideas relating to flutter and aeroelasticity of aerospace vehicles. In the early 1980s, Reed retired from Langley to join the engineering staff of DEI. While at DEI, Reed recognized the need to increase the safety and minimize the cost and hazards associated with aircraft flight flutter testing. He thus conceived and patented the DEI Flutter Exciter, now used world wide in flight flutter testing of new or modified aircraft designs.

The Flutter Exciter system provides dynamic force input to stimulate structural vibration modes of an aircraft in flight, and through all its flying conditions. This innovative product generates programmed sinusoidal forces at controllable frequencies, amplitudes, phases and sweep durations. By inducing these forces, structural vibrations on the aircraft can be damped.

The DEI Flutter Exciter is controlled directly by the pilot/operator through a digital Cockpit Control Box and Avionics Box electronics and software. The Exciters are powered by DC servo motors through interconnecting cables. Small fixed vanes are mounted at the wing tips and tail tips of an aircraft.

When activated, the DEI Flutter Exciter alternately deflects the airstream upward and downward in a rapid manner, creating a force similar to that produced by an oscillating trailing edge flap. Because the system runs on very little electrical power, rather than tapping into an aircraft's hydraulic system, several benefits are realized. Among them, because the DEI flutter components are a "strap-on" type of hardware, cost and complexity of fitting the system on the aircraft can be lessened. The DEI Exciter is readily adaptable to a variety of aircraft. It is effective on all fixed wing and rotorcraft applications, from single-seater to airliner-sized vehicles.

DEI sub-scale flutter exciters are also available to incorporate into wind tunnel flutter models, permitting an aircraft builder to pre-qualify a vehicle design.

In recent years, DEI used its skills to fabricate a complete set of fan blades for the National Transonic Facility (NTF) at Langley. DEI has applied its ideas on suspension techniques to NASA work on large, flexible space structures. More than 1,000 tasks have been completed for Langley alone. Employed by NASA Ames-Dryden, the employee-owned group was also contracted to work on NASA's F-16XL flight test program.

Twenty-five years after creation, DEI has grown to several hundred employees with $20 million in annual sales and has expanded its aerodynamic analysis techniques to justify its company claim: "Where imagination takes shape."

Dynamic Engineering Inc.'s Flutter Exciter mounts at the wing tips and tail tips of an aircraft to damp out structural vibrations.

Intelligent Fasteners

I f you are facing that oft-quoted engineering truism, that a system is only as good as its weakest link, consider the world's first, high-temperature-resistant, "intelligent" fastener.

A product of Ultrafast, Inc. of Malvern, Pennsylvania, this fastening technology was originally developed under a Small Business Innovation Research (SBIR) contract managed by Marshall Space Flight Center. The NASA partnership was born by a need for critical-fastening appraisal and validation of spacecraft segments that are coupled together in space. In-orbit assembly requires both lightweight wrenches for enhanced robot-arm mobility as well as remote fastener-load inspection capability.

The space solution yielded an innovation that is likely to revolutionize manufacturing assembly, particularly in the automobile building industry.

Ultrafast's "intelligent bolt" utilizes a piezoelectric thin-film deposited directly on one end of the fastener. When electrically excited by an Ultrafast tool, tensile loads can be accurately controlled during the bolt tightening process. Insufficient bolt preload is usually the root cause of joint failure resulting from joint separation, bolt loosening or fatigue.

In effect, a bolt topped by the thin-film technology—at a cost of just pennies per fastener—functions as a transducer for measurement and recording of bolt tensile load. The coating itself is less than 0.001 inch thick and is durable, deposited on the fastener by sputtering. This is a vacuum process that has long found practical application in everything from integrated circuits to reflective coatings on glass or decorative coatings on plastic. Moreover, the coating can be applied to all forms of existing fasteners without changing the basic design or metallurgy of the item undergoing the process.

Ultrasonic measurements of a fastener are possible by using piezoelectric thin films, in both longitudinal and transverse directions, heretofore impractical in typical fasteners. Ultrafast technology uses the relationship between the speed of ultrasonic waves in a material and the stress applied to the material as its basis for computing load measurements. The idea is that the "time-of-flight" of an ultrasonic signal traveling in a fastener will increase as the load on the fastener is increased.

Ultrafast's intelligent-bolt technology deletes the self-defeating procedure of having to untighten the fastener, and thus upset the joint, during inspection and maintenance. Even the smallest fastener can be turned into latent sources of information in numerous applications.

For the automotive industry, the advantages of applying the Ultrafast system are distinct. Safety-critical components like powertrains, steering systems and brakes can achieve higher reliability while minimizing service costs and other impacts from joint failure. Use of high speed impact or impulse wrenches to improve joint integrity and inspection can lower costs in automotive body manufacturing by reducing joint assembly times.

Practical use of ultrasonic thin-film fasteners go beyond the realm of automobile, aircraft, and space applications. For instance, from computer disk drive assemblies to forklifts. In the nuclear field, from small flanges to large pressure vessels. For chemical processing, critical-fastening applications include offshore platforms to pipes. In construction, the appeal of Ultrafast's innovation can be applied to buildings and bridges.

Ultrafast expects sales to exceed $100 million in a few years. Ultrafast has licensed two major European tool companies to supply power tools for automotive assembly and is establishing worldwide fastener coating services through its Ultracoat subsidiaries.

Ultrafast's "intelligent" fasteners are expected to be a boon to automobile manufacturers, assuring that more precise loads are applied during bolt tightening.

Automotive Insulation

Beating the heat in a stock car competition is another kind of space race that demands thermal protection materials.

Under a Space Act agreement between Boeing North America (formerly known as Rockwell Space Systems) and BSR Products, Inc., of Mooresville, North Carolina, Space Shuttle Thermal Protection System (TPS) materials that orbit the globe now circle the race track.

BSR has created special TPS blanket insulation kits for use on autos that take part in National Association for Stock Car Auto Racing (NASCAR) events, and other race cars through its nationwide catalog distribution system.

NASA's fleet of orbiters are protected by TPS materials inside and out, from the sometimes space-exposed cargo bay to the space plane's outer surface. Developed by Rockwell, classes of TPS tiles and thermal blankets safeguard Space Shuttle orbiters as they slam back into Earth's atmosphere. Reentry heat loads can be as high as 3,000 degrees Fahrenheit during the plunge.

The idea of using Space Shuttle TPS to insulate heat-generating areas of stock cars came by way of a tour taken by NASCAR champion Bobby Allison at Kennedy Space Center (KSC). Then KSC Director, Jay Honeycutt, a racing fan himself, recommended to Allison that TPS insulation could shield drivers from excessive heat exposure.

Speeding race car drivers are in the hot seat in more ways than one. It has been estimated that temperatures inside a race car's cockpit can soar to a sweltering 140 to 160 degrees. It is common for NASCAR drivers to endure blisters and burns due to the excessive heat flooding into the cockpit. That extreme heat comes through the engine firewall, transmission tunnel, and floor. High temperatures, in fact, have led to many

Butch Stevens of BSR Products, Inc., lays out a custom exhaust thermal protection system in front of the Motorsports Busch Grand National race car. BSR created blanket insulation kits based on NASA Space Shuttle Thermal Protection System materials and had the first products bear a seal from the U.S. Space Foundation indicating their space origin.

BSR Products, Inc.'s line of insulation includes floor panels, oil tank blankets, filters, transmission tunnel blankets, and exhaust crossover shields. The insulation lowers race car temperatures and keeps drivers cooler.

totally exhausted drivers being hauled out of their cars after a grueling meet.

NASCAR racer, Allison, intrigued by the use of TPS, contacted colleague and rival, Roger Penske, who was able on short notice to loan a stock car to KSC for one day of TPS retrofitting.

Penske team members, Rockwell and NASA personnel worked together to pattern TPS material to fit Penske Racing Inc.'s No. 2 Ford Thunderbird stock car. The TPS insulation extras added less than four pounds to the car. Later tests clearly showed significant temperature drops in locations where the TPS material was used. In the driver's cockpit, temperatures were lowered by some 50 degrees.

A big thumbs up was given to the TPS additions by NASCAR driver Rusty Wallace. He raced several times with the material and participated in an instrumented test at Daytona International Speedway in April 1996. "This is a breakthrough," Wallace says. "I am totally impressed with this material. I feel that the TPS material helps the whole car run cooler, and the cooler the car, the better the performance," he says.

Enter BSR. BSR-TPS Products, Inc. is now manufacturing insulation kits for distribution to race car teams around the world. The cost is a modest $1,300 apiece, with the company expecting to generate $1 million a year in sales.

BSR identified numbers of areas in stock cars that would benefit from selective touches or a mix of Space Shuttle TPS materials, such as: under the drivers seat and ancillary components; between floorpan and exhaust system on the drivers side; for insulating the oil tank; to shield the ignition system; and for installation under the driver's feet and along the side of the transmission tunnel and behind the pedals.

Martin Wilson, Rockwell's TPS Facility project manager, says that the match of TPS, the Technology Programs and Commercialization Office at KSC, and NASCAR racing is sure to significantly improve environmental conditions for race car drivers. "This is another good example of how technology developed for the space program can be used for applications on Earth," Wilson says.

Invisible Flame Imaging

As the truism goes, "where there is smoke, there is fire." But with a device called FIRESCAPE™, firefighters can now see invisible flames from alcohol and hydrogen fires during the day and even see through smoke.

An electronic flame imager, originally developed by NASA for rocket engine testing, is being manufactured and sold to firefighters by SafetySCAN, LLC. of Buffalo, New York. SafetySCAN, a company that specializes in opto-electronic sensor and display systems for firefighters, calls FIRESCAPE the first affordable commercial product for invisible (or ashless) fire imaging.

The technology was developed by John C. Stennis Space Center to visually assess the presence, location, and extent of hydrogen fires. The need for such equipment was generated by the center's use of more than one million gallons of liquid hydrogen per month in its rocket engine test programs. Indeed, hydrogen fires are a significant risk.

Previously, firefighters responding to a hydrogen fire had to give the suspect area "the broom test" by carefully probing the suspect area with a corn straw broom to determine the presence and location of a fire. This technique has significant safety and accuracy shortfalls, particularly in windy outdoor conditions where flames can easily change direction.

"There was a huge gap in technology between the $3 broom and the $30,000 thermal imagers," says Heidi Barnes, a Stennis engineer who, along with colleague Harvey S. Smith of Lockheed, originally developed the device. "Firefighters need a reliable but economical device to assist them in their work. The technology was there; it has just been a matter of developing something relatively simple to use and getting it out there to them," Barnes says.

The NASA-developed equipment detects and images the infrared emissions from the final combustion process, providing a true depiction of the fire size. Heat from vapors or surfaces is ignored, while the flames can easily be identified in the context of the unaltered surroundings. Sunlight, fog, smoke, or mist do not significantly diminish imaging capabilities. The hand-held hydrogen fire imager can detect invisible flames of hydrogen and alcohol fires, but is also helpful for firefighters in controlling conventional fires.

As a result of a market assessment performed by Research Triangle Institute (RTI) for Stennis, the fire imaging technology was determined to have commercial benefit. Based on this information, NASA applied for a

NASA Stennis engineers devised a way to scan for hydrogen fires as part of rocket engine test programs. The technology fostered development of FIRESCAPE, a commercial unit to "see" the invisible flames of hydrogen and alcohol fires.

NASA engineer Heidi Barnes demonstrates the antiquated "broom method" of detecting invisible hydrogen and alcohol fires.

The Stennis Space Center fire department illustrates how the fire imager can be used to see through dense smoke to find a staged victim trapped in the building.

patent. NASA's Technology Transfer Office then facilitated a briefing at which companies were introduced to the technology. Based on a review of competing product commercialization business plans, NASA granted exclusive patent rights to SafetySCAN.

SafetySCAN has subsequently packaged the technology into a hand-held device weighing five pounds. FIRESCAPE has no moving parts and is used like a pair of binoculars. The optics are sealed within a case to protect them from smoke and grit. Within a five-second period, the device is up and running and can be used for two hours before recharging. A standard video output jack is provided to connect the imager with an external monitor or video cassette recorder. At a current price of $5,000, FIRESCAPE is within the budget of most fire departments.

In addition to safety, a primary benefit of using the fire imager is the simple employment of the device. With a push button on/off switch, and a button to compensate for sunny and cloudy conditions, the fire imager is a snap to operate.

"SafetySCAN has been aided tremendously through the NASA technology transfer process; not only for the technical content developed into FIRESCAPE, but also for the national exposure associated with the process. This national exposure has resulted in early sales of FIRESCAPE," says Mark Stroze, president of SafetySCAN.

Stroze adds that, ultimately, the fire safety market will be an even bigger winner through the potential of greater loss prevention. "And we are proud to say it all came about through the dedicated efforts of NASA, its affiliates and SafetySCAN personnel," he says.

™ FIRESCAPE is a trademark of SafetySCAN, LLC.

Infrared Camera

A sensitive infrared camera that observes the blazing plumes from Space Shuttle or expendable rocket liftoffs is capable of scanning for fires, monitoring the environment and providing medical imaging.

The hand-held camera uses highly sensitive arrays of infrared photodetectors known as quantum-well infrared photodetectors, better known as QWIPs. Advances in the growth of gallium arsenide semiconductors have led to quantum-well infrared photodetectors sensitive to long-wavelength infrared radiation.

QWIPs are the brainchild of the Jet Propulsion Laboratory's (JPL) Center for Space Microelectronics Technology in partnership with Amber, a Raytheon company in Goleta, California.

A quantum well can be visualized as a small well with electrons in it at a state of rest. When these electrons are displaced by a photon—the smallest energy unit in a beam of light—electrons burst out. The current produced by the electrons is relative to the amount of infrared photon energy which struck them. By measuring that current, the photodetector can tell how much infrared light comes from various sources at the scene being imaged.

Using a visible CCD camera, the Malibu fires are not detectable.

QWIP's set of infrared eyes are sensitive to heat energy in the eight to twelve-micrometer wavelength range. This is 20 times longer in wavelength (lower in energy) than visible light. That allows the camera to see radiation at wavelengths not normally visible to the human eye. Room temperature objects, observed at these wavelengths, can be seen to radiate the same way red-hot objects glow in visible light.

As a portable infrared camera, the 10-pound camera offers unrivaled sensitivity in recording objects that give off wavelengths in an outermost part of the infrared spectrum.

For infrared light detectors to function they must be maintained at cold temperatures. QWIP comes replete with a Stirling cooler, a closed-cycle refrigerator about the size of a fist. Despite its modest proportions, this motor cycles cooling gas millions of times. That processing drops the camera's temperature from room temperature to minus 343 degrees Fahrenheit—in all of 10 minutes.

The eye of the camera is a two-dimensional array of detector pixels, with each pixel converting some of the infrared photons to an electric signal. That array consists of 65,536 QWIPs to make up a focal plane of 256 by 256 pixel. Each QWIP is a pixel. Size wise, the entire camera measures 4.4 inches wide, 10.3 inches deep and 7.2 inches long. It can be plugged into a video cassette recorder.

Due to the camera's industry/government genesis and its extraordinary capabilities, QWIPs are ideal for a variety of ground and space-based applications. Night vision, early warning systems, navigation, flight control systems, weather monitoring, security and surveillance are among the duties for which the camera is suited. Its sensitivity to the distribution of different gases also adds to QWIP's role as an environmental watchdog to measure pollution and relative humidity profiles in the atmosphere. It works effectively in both daylight and nighttime conditions.

Medical applications are also expected. One QWIP camera underwent testing by surgeons at the Texas Heart Institute to see which arteries are carrying blood during heart surgery.

With the QWIP camera, users could see through smoke and pinpoint hotspots at the Malibu fires in October 1996.

In late October 1996, the QWIP camera debuted as a fire-observing device. The camera hopped a news helicopter flight, courtesy of KCAL-TV, Channel 9 in Los Angeles, California. A spate of destructive fires were speeding through Malibu. JPL's Mani Sudaram, a member of the QWIP development team, used the camera to transmit live images from the helicopter. High-flying QWIPs were quick to point out hot spots in areas that appeared innocuous to the naked eye. These sites, easily discounted as hazardous on first view, are of concern for firefighters as they can flare up even after the fire appears to have subsided.

From tumor detection and environmental monitoring to military applications and law enforcement, this advanced infrared camera shows great promise.

The hand-held QWIP camera measures infrared light in applications ranging from fire scanning to medical imaging.

Sensor Validation Software

Call it "reasoning with uncertainty." That is the raison d'etre of Expert Microsystems, Inc. of Orangevale, California.

The company developed SureSense™, real-time sensor data validation software. The work was a direct result of a Small Business Innovation Research (SBIR) contract with Lewis Research Center. This ultra-reliable control and sensing system product was produced and distributed through a partnership in 1994 between Expert Microsystems and Intelligent Software Associates, Inc. (ISAI).

SureSense was created in response to a NASA need for verifying the reliability of sensor input that operates advanced automation and control systems.

The SureSense software had an immediate assignment: improve safety and reliability of Space Shuttle Main Engine (SSME) operations. Along with its unprecedented number of successful flights, Space Shuttle history also illustrates the potential value of sensor validation. The program has suffered test aborts, launch delays and postponements as a result of sensor failures. Dangerous conditions may result when the sensors used to prevent catastrophic system malfunction fail themselves.

SureSense was designed to detect failures in sensors that must function in a trustworthy manner. Erroneous engine shutdowns due to sensor failures are costly, with any test abort or launch delay forcing an expensive and time-consuming anomaly investigation.

In one dramatic instance, sensor failure caused the premature in-flight shut down of an SSME roaring toward orbit. Space Shuttle mission 51-F in July 1985 had the number one SSME shut down early. Luckily, the

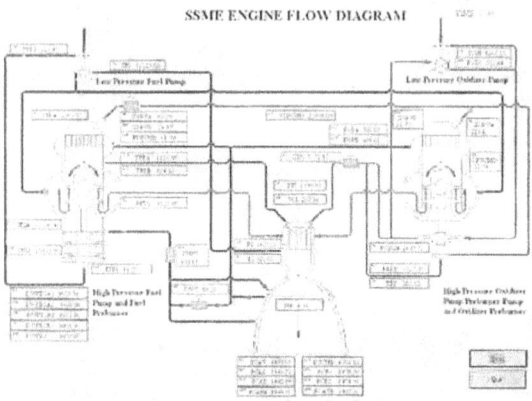

Expert Microsystems, Inc.'s validation software was developed for verifying sensor operations in Space Shuttle Main Engines (SSMEs). The user interface screen shows normal operation of the SSME during testing at Marshall Space Flight Center's avionics simulation facility.

Shuttle was over five minutes into its ascent and was capable of an "abort to orbit." On the second engine, sensors also indicated that shutdown was necessary. A mission operations engineer noticed an incongruency in the sensor readouts and advised against it.

Expert Microsystems's SureSense real-time sensor validation software reduces the workload, schedule and uncertainty associated with sensor failure identification and recovery. Furthermore, data integrity also improves

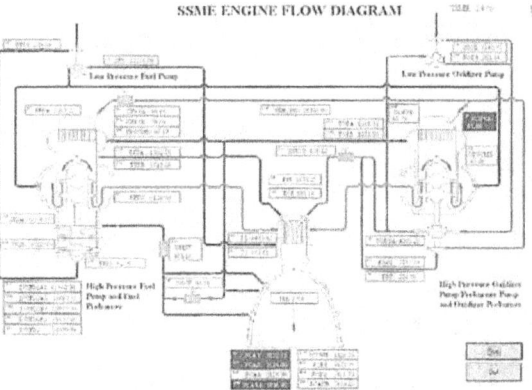

In real-time, SureSense software can depict failure of several mission critical sensors on the Space Shuttle Main Engine and highlight the sensor value display in red.

safety, dependability and economics for any mission critical aerospace or industrial control application.

Sensor validation is advantageous for several reasons. Among them:

- averts safety system false alarms that leads to unwarranted shutdown or maintenance,
- lengthens hardware life and assures mission success,
- provides reliable "red-line" safety protection for personnel and equipment, and
- strengthens system automation by ensuring that automated checkout and diagnostic systems "reason" with valid data.

The NASA work led to recognizing the wide-spread commercial potential of sensor data validation software, says Randy Bickford, President of Expert Microsystems. The company has structured the software to enable application to virtually any process control environment, such as computer integrated manufacturing, power plants, and hazardous gas sensing and control systems.

It is estimated that the nationwide market for special purpose industrial controls is in the billions of dollars, indicating the size and diversity of the process controls market.

™ SureSense is a trademark of Expert Microsystems, Inc.

Portable Radiation Detectors

An advancement in more portable radiation detectors was made possible by satisfying a NASA need for a non-clogging Joule-Thomson (J-T) cryostat to provide very low temperature cooling for various sensors.

Through a Small Business Innovative Research (SBIR) contract from Kennedy Space Center, General Pneumatics Corporation's Western Research Center (GP WRC) in Phoenix, Arizona was tasked to overcome reliability problems and other shortfalls in conventional J-T cryostats. Later, GP WRC worked with EG&G ORTEC in Oak Ridge, Tennessee, to apply the technology to cooling high-resolution gamma-ray spectrometers, under a contract to Bechtel Nevada. The U.S. Department of Energy's Remote Sensing Laboratory in Nevada, operated by Bechtel Nevada, directed the development.

J-T cryostats produce cryorefrigeration by expanding a gas from high pressure through a nozzle. In many uses, the cryostat automatically regulates the flow to provide rapid cooldown and then conform to match the heat load. A common problem in conventional cryostats is that the very small diameter orifice of the nozzle can easily become clogged by contaminants in the flow.

In the NASA SBIR project, GP WRC developed an anti-clogging, flow-regulating J-T cryostat. The GP WRC cryostat design has a tapered annular expansion nozzle which is highly resistant to clogging due to a large ratio of circumference to flow area. Grooves in the nozzle increase flow friction to allow wider, less clog-susceptible passage. Turbulent voids continuously break up and clear contaminants from the flow. The GP WRC J-T cryostat design allows several hours of continuous operation with gas contamination levels that would clog conventional cryostats within six minutes.

GP WRC's J-T cryostat also solved flow regulation difficulties in cooling the gamma-ray detectors. Flow regulation is induced by the heat content of the return flow gas, instead of by the temperature near the nozzle as in conventional cryostats. This provides a large flow for fast cooldown, while conserving gas according to the heat load at operating temperatures.

This NASA-supported cryostat development played a key part in the development of more portable high-purity germanium gamma-ray detectors. Such are necessary to discern between radionuclides in medical, fuel, weapon, and waste materials. Despite the abatement of the Cold War, the ability to monitor nuclear materials, verify possible hazards, and develop counter proliferation tactics has become increasingly crucial to global security.

A high-purity germanium gamma-ray detector is typically cooled with liquid nitrogen, which may take 10 hours or more because of the relatively large, thermally-isolated mass. The dependence on liquid nitrogen severely limits the mobility and continuous operating time of such detectors. Electric-powered mechanical cryocoolers induce unwanted vibration or are too bulky and power-consuming to be very portable. Thermoelectric coolers, while vibrationless, cannot achieve the necessary temperature or power efficiency.

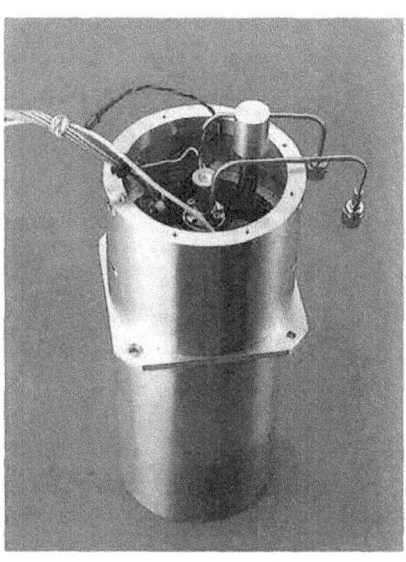

J-T cryostats produce cryorefrigeration by expanding a gas from high pressure through a nozzle.

Essential to developing a more portable gamma-ray detector unit was the J-T cryostat expertise gained by General Pneumatics as a result of its NASA SBIR project. The outcome was a cryostat that can cool gamma-ray detectors, without vibration, using compressed gas that can be stored compactly and indefinitely in a standby mode.

In addition to the gamma-ray detector application, General Pneumatics has parlayed its NASA-backed research to produce custom J-T cryostats for other government, commercial, and medical applications.

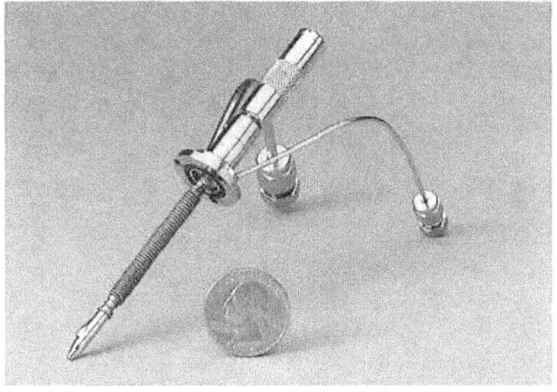

General Pneumatics cryostat nozzle work was pivotal in building more portable gamma-ray detectors. A prototype J-T cryostat-cooled HPGe high-resolution gamma-ray detector is designed to monitor nuclear materials.

Stress Management by Biofeedback

Exercise and stress management programs designed for high-flying astronauts are just the ticket to help reduce on-the-job anxiety and hypertension, and to find calm in a harried workplace. Bio-Games© were formulated in the 1980s when psychologist, Patrick Doyle, served on a project to train U.S. astronauts at Johnson Space Center in biofeedback techniques. Traditional biofeedback concepts—which can be merely listening to a tone and watching the body respond on a graph—were found to be too mundane, repetitive and boring.

Doyle's response was to develop more interesting and involved formats. As a first product, Bio-Ball™ was established as an interactive, multimedia baseball video game that is played by relaxing in order to hit the ball. Gradually, the player learns to relax at will, and with practice, is soon able to generalize the newly acquired skills to real-life situations.

Bio-Ball was well received by NASA, with Doyle moving on to other high interest format software that can manage stress. As an associate professor of psychology at the University of Houston/Clear Lake, Doyle has gone on to create a number of biofeedback games.

A Bio-Games series of interactive video products is now being marketed by the Houston-based Creative MultiMedia, Inc. (CMM). CMM has issued such games as: Bio-Ball©, Bio-Golf©, Clutch City©, 3-D Space Pilot©, and Pachyderm©. Three inexpensive EMG sensors are connected to the players' forearms or other muscle sites, monitor tension levels, and communicate that information to the game. The sensors allow a person to play the games without touching their computer keyboards. The multimedia Bio-Games run on any 486 or better DOS-based computer with Windows.

Astronaut relaxation techniques that use biofeedback have fostered stress-relieving commercial products. Bio-Ball is a baseball game in which deep muscle relaxation controls events rather than joystick manipulation.

In Creative MultiMedia, Inc.'s Bio-Golf, nine holes are played only by learning to discriminate and create different levels of tension.

In Bio-Golf, for instance, the players register three tension levels: high tension for driving; middle tension for chipping; and low tension, or relaxed, for putting. The player then has to recall those levels to make the appropriate shot. A too-tense shot on the computerized golf course causes the ball to overshoot. Too relaxed means the ball falls into the water or sand.

"The computer games offer stress management in an enjoyable format so that players can learn to handle life events effectively. As players interact with Bio-Games, there are continuing challenges and new levels of difficulty to keep interest high," Doyle says.

Dr. Doyle explains, "Several of the Bio-Games build team skills. Players are teammates against the computer. The players feel engaged and excited in a competitive sense, but must work together to overcome the challenges and stay calm under pressure."

Stress-busting screen savers are also being marketed by CMM, under the Buddies series. Buddies involves two animated characters that assist an employee to implement stress management intervention through deep breathing, progressive muscle relaxation, imagery, and cue word exercises. Another screensaver is Meditation. This product takes the deep breathing and concentration techniques of yoga and mixes it with computer technology.

Doyle's work and the CMM products have been recognized by Steven Spielberg's Starbright Foundation which focuses on improving the total hospital environments of critically injured and chronically-ill children.

More products are being planned. From astronaut to overworked executive, the biofeedback stress-management tools offer a new way to nurture safety and health, as well as enhance productivity, in the frantic society of today.

Music Magic © is an interactive piano keyboard that responds to different levels of deep muscle relaxation.

Ergonomic Chairs

The posture that an astronaut's body assumes while floating in the microgravity of space has provided anthropometric data useful in designing chairs for the workplace.

Measurements recorded in orbit by NASA astronauts have shown, in its most unstressed and relaxed state, the human body assumes a trunk-to-thigh angle of 128 degrees. This zero-gravity posture was found to place the musculoskeletal system in its most rested condition. Specifically, this natural posture fosters a non-stressed muscle system, correctly aligned vertebrae, better breathing, improved digestion, and enhanced circulation.

Crew members living aboard NASA's roomy Skylab orbital station in the mid-1970s noted that in space, the human posture differs from the normal posture caused by the tug of one gravity. These findings and others were collected by the Johnson Space Center, then published in a NASA Anthropometric Source Book.

This reference material helped BodyBilt®, Inc. of Navasota, Texas, a subsidiary of ErgoBilt, Inc., to fashion controlled comfort chairs that lessen the harmful effects of gravity on seated workers. A brochure from the company puts the product into perspective: "A pressure-reducing contour that lessens the devastating effects of gravity. From space-age research, a seating system developed to support a natural, stress-free posture."

BodyBilt's ergonomically-correct line of office chairs are targeted for the average worker that sits for prolonged periods, be it in the cleanroom or the boardroom. There has been an alarming increase in back pain and muscle fatigue in workers, along with a dramatic escalation in repetitive stress injuries. In fact, U.S. Department of Labor statistics show that more than 60 percent of all occupational illnesses are now due to repetitive motion disorders, including those resulting from sitting and carrying out intensive tasks. The costs of claims by employees afflicted by these problems is on the rise. Lower back pain, exacerbated by poor seated posture and inadequate weight support, is second only to the common cold as a cause for missed work. Back problems are the number one injury for workers under the age of 45.

Crafted into BodyBilt chairs is a 10-point posture control system and pressure-reducing seat contours. These features, among others, are designed to reduce work injuries while increasing comfort and productivity.

BodyBilt's Task Series is designed for users who require a high degree of lower back support for desktop-intensive tasks. To meet a variety of workplace needs, BodyBilt also offers a Management Series chair for managers and executives. Also, one group of workers—those above-average built employees—are served by BodyBilt's Big & Tall Series chair, which is designed specifically to accommodate users up to 350-lbs. The company also offers a line of medical chairs and a new line of side chairs (guest seating).

The BodyBilt line includes a range of features, such as an adjustable backrest, pivoting seat arms, and tilt tension control. A specialized lumbar support structure can be augmented by the optional Air Lumbar™ Pump, giving each user a custom fit for highly effective lower back support. One-touch controls give the chair user quick access to the natural, stress-free posture that mimics the absence of gravity.

BodyBilt's roster of national clients lists such organizations as IBM, Microsoft, Texas Instruments, Hewlett Packard, Eastman-Kodak, Boeing, Motorola and Walt Disney Studios.

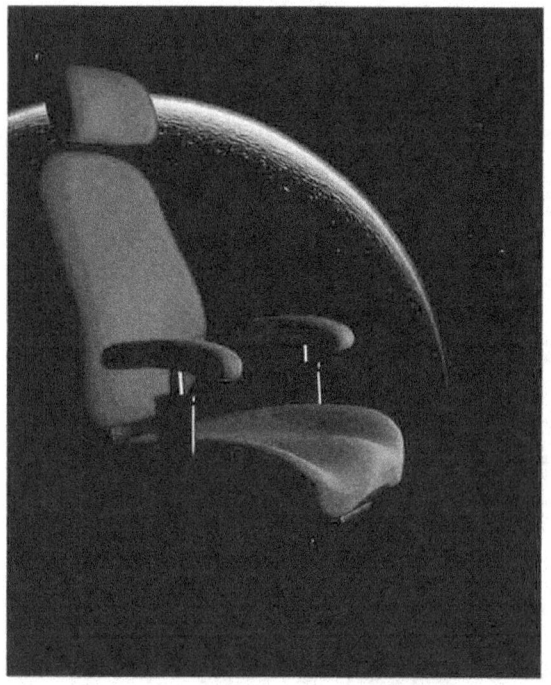

Studies of astronauts floating in microgravity have yielded data useful in the design of chairs that relieve the body of back and leg pain, as well as muscle fatigue in the workplace.

® BodyBilt is a registered trademark of BodyBilt, Inc.

™ Air Lumbar is a trademark of BodyBilt, Inc.

Toy Gliders

Toymakers at Hasbro, Inc. of Pawtucket, Rhode Island are delighted with the "ups and downs" of a product that benefited from NASA expertise. Toy designers at the firm set their sights on creating a foam glider, one that a child could fly with little knowledge of aeronautics.

But early in its development, the Aero Nerf® gliders had one critical problem: they didn't fly so well! For toy gliders, that can surely dampen commercial sales.

Through the expertise of Don Steinman of NASA's Northeast Regional Technology Transfer Center, members of a Hasbro team were linked with aeronautical experts at Langley Research Center in Virginia. As glider hobbyists themselves, the NASA engineers provided essential information about how wing design and shape are integral to a glider's performance.

At issue was improving the overall flying distances and the loop-to-loop stunt qualities of Hasbro's prototype gliders. Langley's Gaudy Bezos-O'Connor, an experimental aerodynamicist by training and a technology transfer specialist, took on the challenge. So too did Ray Whipple, a wind tunnel manager at Langley and specialist in testing model airplanes. Two retirees from Langley, Dave Robelin and Hewitt Phillips, also brought their aeronautical know-how to the project, rounding out the NASA team.

The benefit of the NASA and Hasbro partnership was indeed an uplifting experience, in more ways than one. The toy company is intent on becoming an aggressive force in the toy glider market, part of the highly competitive, multi-billion dollar toy industry. As a step toward that objective, Hasbro's senior design director, Todd W. Wise, perhaps put it best, "Who knows better how to make things fly than NASA."

Elegant but simple was the rule of the day. The gliders had to fly and perform on the basis of their shape, without benefit of changes in any control surface. Propulsion for the glider had to come from a toss of the hand, without the aid of a sling shot or other device.

The Hasbro designers received from NASA not only technical guidance but a hands-on tutorial on the physics of designing and flying gliders. Where to place the wings on a glider's fuselage and the correct angle for its tail surfaces proved critical in the foam toy's make-over. With a better understanding of flight and how to shape the foam, Hasbro designers went back to their drawing boards. The final products, quite literally, "soared" beyond the design team's expectations.

The result? Several versions of a Nerf glider were realized from the collaboration. For instance, the Super Soaring Glider™ can make long-range, high-performance flights. The Ultra Stunt Glider™ is ideal for performing aerial acrobatics. It can execute thrilling loops and turns, all from a simple throw.

NASA has long shared its imaginative approaches in designing aircraft, teaming with aviation and aerospace companies. The Aero Nerf gliders meant leveraging decades of aeronautical skill honed at Langley in scale-model, low-speed, aircraft design research and wind tunnel testing.

A real toy story come true!

® Aero Nerf is a registered trademark of Hasbro, Inc.
™ Super Soaring Glider and Ultra Stunt Glider are trademarks of Hasbro, Inc.

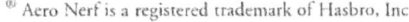

Winging their way into toy stores are Hasbro Aero Nerf Gliders, benefiting from NASA wind tunnel and aerodynamic expertise.

Spherical Camera

Everyone likes to get the complete picture. But for Interactive Pictures Corporation (IPC) of Knoxville, Tennessee the whole scene is a panoramic 360-degrees and looking at four views at once.

The interactive imaging technology, called IPIX™, was developed largely as a result of a Small Business Innovation Research (SBIR) contract through Langley Research Center. NASA found the technology appropriate for use in guiding space robots, in the space shuttle and space station programs, as well as research in cryogenic wind tunnels and for remote docking of spacecraft.

Based on NASA Small Business Innovation Research contracts, interactive imaging technology by IPIX combines two images from 180° perspectives into a sphere. Dr. Lee Martin of IPIX captures the images digitally.

An innovative imaging process, IPIX provides for real-time control of live video data. Viewers can look in one direction from a single vantage point or four views simultaneously.

IPIX technology relies on digital image manipulation of a standard video signal. This permits independent panning, tilting, zooming and rotating of four video images. It requires no moving parts, is noiseless, and responds faster than the blink of an eye. Key elements are a high resolution charge coupled device (CCD), image correction circuitry and a microcomputer for image processing.

A further development of the IPIX technology is spherical photography. A digital camera or standard 35mm camera fitted to a fisheye lens is required. Images taken are processed and electronically scanned to create a digital file. Once digitized, a single fisheye shot captures a hemisphere, while two opposing shots provide an entire sphere. Using seaming tools, the two opposing shots are fused without a discernible seam.

Images of any location are captured in their entirety in a 360-degree immersive digital representation. The viewer is positioned inside the 360-degree digital image and can navigate to any desired direction within the image via a computer mouse, joystick, or other input device. Any part of the image can be enlarged for detailed exploration.

Potential applications of IPIX technology include the viewing of homes for sale, hotel accommodations, museum sites, news events, and sport stadiums. Several car manufacturers already use IPIX to give a viewer a behind-the-steering wheel look at their latest line up of automobiles.

Uses of IPIX technology are far reaching. One application is for non-invasive surgeries. By implementing OmniScope™, also developed by IPC, surgeons can look more closely at various parts of an organ with medical viewing instruments now in use.

OmniScope allows the surgeon to see more without repositioning the camera all the time, says Daniel Kuban, chief operating officer of IPC. "The OmniScope becomes an extension of the surgeon's eyes. It frees the surgeon to focus on the operation," Kuban comments.

An OmniScope benefit to the surgeon is post-operation evaluation of any aspect of the surgical area in detail because a wide-angle, digital image has been captured.

IPIX technology means that tiny cameras that can see a whole room, could revolutionize the security and surveillance industry. In addition, teleconferencing, virtual reality and telepresence theme parks, and a host of military operations are among the applications for this imaginative technology.

™ IPIX and OmniScope are trademarks of Interactive Pictures Corporation.

Once the two images are joined, users can select sections of the sphere and zoom in on desired areas. The technology is finding applications in real estate sales, surgery and museum sites.

Force Feedback Joystick

Any computer game joystick jockey will tell you: It's all in the touch. But a new technology called Force Feedback is adding vivid physical "feel" sensations to computer entertainment systems.

"Get a grip" takes on added dimension in the fast-paced, virtual world of computer gaming. Experience the recoil from shooting a gun in virtual reality. Perceive the jarring bounce, bump, and vibration while driving over simulated road terrain. Feel the reaction of your computerized dragster as it slams into the race track wall.

I-FORCE™ is a computer peripheral from Immersion Corporation of San Jose, California. This Silicon Valley-situated high-technology firm has taken interaction to a new dimension—one that literally comes alive in your hands.

I-FORCE was derived from virtual environment and human factors research and done at the Advanced Displays and Spatial Perception Laboratory at Ames Research Center, in collaboration with Stanford University's Center for Design Research.

Entrepreneur Louis Rosenberg, a former Stanford researcher, now president of Immersion Corporation,

An early Force Feedback prototype was derived from work done at NASA's Advanced Displays and Spatial Perception Laboratory.

credits much of the knowledge acquired to move Force Feedback into the commercial world to the NASA-university joint research efforts. Rosenberg collaborated with Dr. Bernard Adelstein at Ames on studies of perception in virtual reality. Immersion Corporation adapted the basic qualities of Force Feedback: the crispness of initial contact, the hardness of surface rigidity, and the cleanness of final release, all in a virtual environment.

"It took us three years and much innovation, but Immersion can now produce consumer Force Feedback

products," Rosenberg says. The result was an inexpensive way to incorporate motors and a sophisticated microprocessor into joysticks and other game controllers. These devices can emulate the feel of a car on the skid, a crashing plane, the bounce of a ball, compressed springs, or other physical phenomenon.

Immersion anticipates that the technology has the potential to increase the realism of the game environment and provide a more engaging and entertaining experience. By adding physical sensations to the toolbox of perceptual effects that game developers have at their disposal, Immersion contends that new life can be breathed into common video game paradigms. In essence, may the force be with you.

"Force Feedback will make the abstract world of software tangible, physical, and personal. It's not just visual any more!" exclaims promotional material from the company.

Immersion Corporation has commercialized Force Feedback technology and added the sensation of feel to joysticks used in computer simulation games. NASA's work in advanced display technologies played a centerpiece in the concept.

To run I-FORCE, Immersion developed a Force Feedback software protocol known as the I-FORCE API. The I-FORCE hardware and electronics architecture has been licensed to numerous manufacturers of computing peripherals. The first products incorporating the I-FORCE technology appeared in 1996. For example, CH-Products of Vista, California embodied the I-FORCE technology into their popular line of FlightStick and CombatStick controllers. Many more products from various vendors using I-Force are to follow as major hardware manufacturers have endorsed the Force Feedback concept.

Immersion's Rosenberg predicts that Force Feedback may go far beyond joysticks, such as a tool for handicapped people to escort unsteady hands or literally pushing open doors during a cyber-tour of a house for sale. Already, Immersion offers the Virtual Laparoscopic Interface™ that can hone the skills of a surgeon through realistic surgical simulations. In a similar fashion, the delicate nature of catheter-based procedures are also being targeted as an emerging simulation.

™ I-FORCE and Virtual Laparoscopic Interface are trademarks of Immersion Corporation.

Thermal Clothing

How best to lessen the prospect of having your toes and fingers freeze in winter conditions or keep you from that one last ski run? Gateway Technologies, Inc. of Boulder, Colorado is marketing and developing textile insulation technology that can keep a sunny smile even in bone-chilling climes.

Licensed to Gateway by Triangle Research and Development Corporation (TRDC) of Raleigh, North Carolina, the enhanced thermal insulation stems from Small Business Innovation Research (SBIR) agreements with NASA's Johnson Space Center and the U.S. Air Force. The effectiveness of the insulation stems from microencapsulated phase-change materials (microPCMs) originally made to keep warm the gloved hands of space-strolling astronauts. The materials were considered ideal as a glove liner, to help thwart temperature extremes of the space environment.

NASA-sponsored work on gloves that can keep an astronaut's hands from freezing while working in space has been applied to ski gear.

"Our technology has the capability of making products thinner, of making people or objects warmer longer, or of keeping people cooler," says Gateway president Ed Payne. Founded in 1990, Gateway has further developed the innovative thermal regulating system after obtaining exclusive license to the technology developed for NASA that same year, Payne says.

MicroPCMs, when applied to textile fibers or textile substrates, the rate of heat transfer is slowed and can boost heat capacity by 1,000 percent. That enhanced thermal characteristic is made possible in manmade fiber by adding microPCMs to a chemical solution or polymer prior to fiber extrusion. In the process, microPCMs are integrated inside the fiber itself. In essence, by applying microPCMs, "smart fabrics" can be made.

The applications for the product appear great, from outer wear, housing insulation, blankets, as well as protective firefighting gear and scuba diving suits. Gateway has developed and has begun marketing

thermal regulating products under the trademark, OUTLAST. Bringing the insulation to the garment and textile industries is part of Gateway's strategic marketing plan.

Products made from OUTLAST, such as ski parkas, hunting jackets and thermal underwear, is an approach far different than traditional insulation methods that rely on trapping air within fiber or fabric. Thanks to microPCMs, heat is absorbed causing the microencapsulated material to change phase (from solid to liquid) at the molecular level, storing or releasing heat in response to temperatures next to the skin. By efficiently holding on to heat and disbursing it evenly within the fabric yields a thinner, but more dynamic and protective thermal barrier between the body and the exterior environment. This, in turn, eliminates the need for garments that are awkward and unwieldy for their wearer.

Several products using OUTLAST have already made their way to the activewear marketplace. For instance, Gateway Technologies has granted Boston-based Tempo Shain Corporation exclusive license to feature its OUTLAST in boot and shoe liners.

Gateway has also granted an exclusive agreement with Bula, Inc. to use OUTLAST in winter headgear, hats and caps for hunting and other outdoor sports. Also, C.F. Ploucquet GmbH & Company in Germany has entered into a licensing agreement with Gateway to produce fabrics using OUTLAST throughout Europe.

Yet another example is the Grandoe Corporation who is developing prototype OUTLAST fabrics for use in a variety of men's and women's ski gloves.

From space gloves to ski gloves, microencapsulated phase change material has become an innovation you can easily get your hands on.

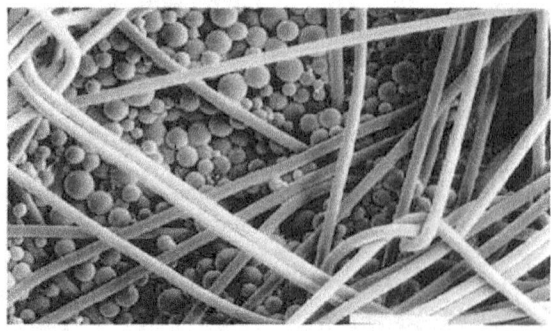

An electromicrograph shows the microencapsulated phase change materials in the insulation for cold-thwarting ski apparel.

Sporting Goods Lubricants

What crawls at one mile per hour, weighs six million pounds, and is always hungering for a lube job? That may sound like something out of a science fiction film. But in truth, this beast is the huge Mobile Launch Platform, needed to haul the fully assembled Space Shuttle from the Kennedy Space Center Vehicle Assembly Building to a launch pad.

In 1994, Lockheed Martin Space Operations was searching for an environmentally-safe lubricant for the lumbering, gigantic transporter. Sun Coast Chemicals of Daytona, Inc., Flagler Beach, Florida was contracted to formulate a spray lubricant free of environmental drawbacks. Demanding specifications were placed on the lubricant, along with the task of providing the crawler with the utmost in lubrication safeguarding.

Working with Lockheed Martin, Sun Coast Chemicals of Daytona (SCCD) was victorious in devising a Crawler Track Lubricant. From this work, SCCD created and introduced three spinoff products: Train Track Lubricant, Penetrating Spray Lube for rust prevention and other related problems, and Biodegradable Hydraulic Fluid.

SCCD has moved forward with yet another two new products, sparked by the original lubricant work.

The X-1R® Super Gun Cleaner and Lubricant (SGCL) has been created for the gun devotee. This product impregnates the gun's barrel and trigger mechanism with microscopic metal pores. Just three to five microns in size, these pores help protect against galling while increasing the velocity action of the gun. A reduction in the heat build-up from rapid fire is also realized.

Designed to clean and lubricate in one step, the X-1R SGCL replaces solvent and oil now commonly used by gun owners. No particulates, such as Teflon® or graphite, are carried in the gun cleaner and lubricant. The environmentally safe, non-hazardous, non-flammable product is fortified with IR-39, a special SCCD advanced rust inhibitor. Making use of the X-1R SGCL protects a gun from rust and corrosion caused by extreme environmental conditions.

Sun Coast Chemicals of Daytona has announced another new product, an offspring from their NASA Crawler Track lubricant work. For the fishing enthusiast, SCCD is marketing the X-1R Tackle Pack.

Due to the high cost of fresh and saltwater reels today, preventative maintenance is a must. With the angler in mind, Sun Coast has developed a premium preventative maintenance package to assure optimum lubrication protection against rust and corrosion. The X-1R Tackle Pack is a blend of SCCD's special Rod and Reel Lube and All Reel Grease. This powerful twosome is a preparation that penetrates, cleans, reduces friction, lubricates, and provides extra protection against rust and corrosion. An angler can cast further with smoother retrievals thanks to the X-1R Tackle Pack.

Once again, the X-1R Tackle Pack is environmentally harmless, non-hazardous, non-flammable, and is used and endorsed by both fresh and saltwater guides and certain reel manufacturers.

Sun Coast Chemicals of Daytona has entered the two new products into retail and wholesale sporting goods markets through the U.S. west coast-based Oshman's Sporting Goods and Academy Sporting Goods chains. SCCD has received its vendor acceptance from the Wal-Mart Super Store chain, as they too have expressed an interest in the SCCD environmentally safe lubricant product line.

Not bad for a company that started out "crawling" just a few years ago, now to become a fast-paced producer of a broad family of environmentally-friendly products.

® X-1R is a registered trademark of Sun Coast Chemicals of Daytona, Inc.
® Teflon is a registered trademark of E. I. du Pont de Nemours and Company.

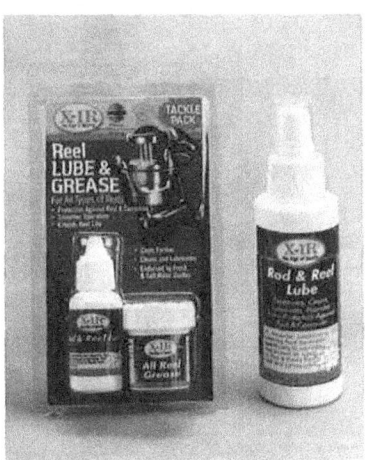

The Reel Lube & Grease is an environmentally safe cleaner and lubricator.

For the hunting enthusiast, the biodegradable Gun Lube Cleaner & Grease protects against corrosion.

Memory Golf Clubs

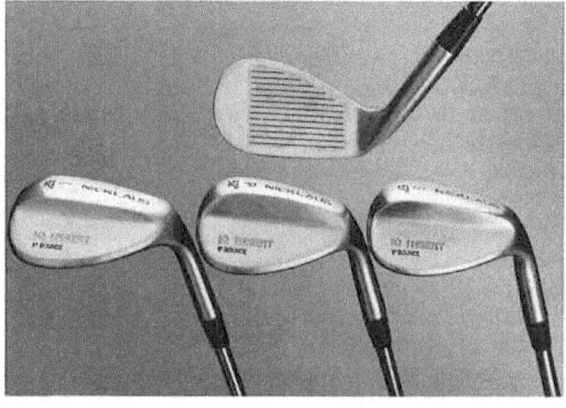

Being in the swing of things is only one consequence of products using technology commercialized by Memry® Corporation of Brookfield, Connecticut. Incorporating proprietary shape memory alloys, a new line of golf wedges and putters has been developed.

Memry Corporation's investigation of shape memory effect (SME) stems from Marshall Space Flight Center contracts to study materials for the space station. SME is a property of certain metal alloys that can change from shape to shape depending on temperature fluctuations. Starting in the late 1980s and early 1990s, Memry made use of its NASA-spurred expertise to create a line of home and industrial safety products.

A proprietary, high-damping shape memory alloy called Zeemet® has been developed by Memry specifically for the golf industry. The Nicklaus Golf Equipment Company of West Palm Beach, Florida has created a new line of golf clubs using Zeemet inserts.

The value of applying shape memory metal into golf club inserts is twofold: its superelastic and high damping attributes. For the avid golfer, that translates into more spin on the ball, greater control and a solid feel.

Upon closer examination, the club insert basically undergoes a split second change in its metallurgical structure upon impact with the golf ball. This elastic property keeps the ball on the club face longer. That supplies more spin to the ball but not at the expense of distance. More "bite" on the fairway is every golfer's desire.

The science of shape memory has permitted Memry Corporation to generate an assortment of commercial sales.

For instance, McDonnell Douglas of St. Louis, Missouri and Memry signed a contract in February 1996 to develop a unique control surface for helicopter blades. Memry is designing and manufacturing small microprocessor-controlled tabs for the trailing edges of the helicopter blades. By controlling the tabs, a pilot can fine tune each blade for improved performance and reduced vibration. The actuation that provides the tab motion is a shape memory alloy torsion device.

Properties of metal alloys studied for the Space Station program have sparked a new line of golf clubs. Shape memory metal gives the most seasoned golfer new control and feel.

Memry's Super-Elastic Nitinol materials have many application within the medical equipment industry. As a binary nickel-titanium alloy, Nitinol can accommodate large strain deformations and spring back to their original shape when stress is removed. This element, therefore, is suitable for catheter guide wires, lesion localizers, suture anchors and dental arch wires.

Other Memry products already available to the consumer include MemrySafe®, an anti-scald product; UltraValve®, a temperature selection and control system for baths and showers; and FireChek®, a reusable fire safety valve for industrial facilities.

MemrySafe is built around the ability to instantly restrict water flow in shower or sinks before scalding. Comprising a memory metal alloy, a patented valve reacts to temperature, not pressure. The unit can sense hazardous scalding water, reducing the flow of hot water to a trickle. When the scalding temperature drops to a less dangerous level, the unit automatically resumes normal flow.

UltraValve is a touch-of-the-button bathing feature. With the press of a button, water can be turned on or off, with water temperature selected according to personal preference. Exact water temperature is displayed at every moment on a user-friendly control panel. This "smart bathing" system also protects its user from scalding water.

FireChek provides for emergency shut down of process control lines that handle flammable and toxic fluids and gases. This product can be applied to any pneumatically-operated process value. If excessive heat is sensed by the shape memory element, the system immediately vents the pneumatic actuator pressure. This closes the supply line. FireChek is suitable for process industries such as petrochemical, chemical, semiconductor, pharmaceutical and large oil and gas fired boilers.

® Memry, MemrySafe, UltraValve, FireChek, and Zeemet are registered trademarks of Memry Corporation.

Jack Nicklaus insert wedges using shape memory metal offers vibration damping qualities that make a real difference on the golf course.

Improving Vacuum Cleaners

It doesn't take a rocket scientist to appreciate how a vacuum cleaner operates. But a touch of space-age engineering has made a clean sweep of the inner-workings of the house-cleaning appliance, making it far quieter and more efficient.

Under a Space Act Agreement between the Cleveland-based Kirby company and Lewis Research Center, NASA technology was applied to a commercial vacuum cleaner product line.

Kirby engineers were keenly interested in advanced operational concepts, such as particle flow behavior and vibration, critical factors to improve vacuum cleaner performance.

Of particular importance to the company was a high-tech evaluation of the firm's 1994 home care system, the Kirby G4™, the results of which contributed to the refinement of the new G5™ and future models.

Under the cooperative agreement, Kirby also had access to Lewis' holography equipment. This apparatus is normally used to analyze the vibration modes of jet engine fans. Laser beams of light can detect vibrations that cannot be discerned with the unaided eye. Using the laser, insight was gained into how long a vacuum cleaner's fan would perform.

Lewis proficiency in advanced computer software that can simulate the flow of air through fans was made accessible to Kirby engineers. Computational fluid dynamics—virtually an "electronic wind tunnel"—was employed to figure out what happens when air or any other substance flows through a passage such as a tube or fan.

The Lewis/Kirby collaboration resulted in several successes, such as fan blade redesign. The new blade was constructed from a polymer that was then configured for a substantial reduction in centrifugal force. Vacuum cleaner blades can run as high as 18,000 spins per minute, compared to just 7,000 to 8,000 in a jet engine. Higher spin rates translates into more stress on the blade. The blade redesign was 300 percent to 400 percent stronger than the previous blade used. A 75 percent noise-level reduction in certain frequencies was also accomplished.

Put into motion was a continuing dialogue between Kirby and Lewis engineers on improving air-flow traits in various nozzle designs. The overall goal is to quantify both velocity fields and particle trajectories throughout the vacuum cleaner nozzle. That information, in turn, can optimize nozzle performance in terms of "cleanability" or ability to remove embedded dirt and other particulates from carpeting or hard surfaces. Any future findings would be incorporated into Kirby's G5 and upcoming models.

The relationship forged between Kirby and NASA was striking. The merger of knowledge and need helped an American company become more competitive in the global market by the sharing of world-class knowledge and state-of-the-art equipment.

™ G4 and G5 are trademarks of Kirby company.

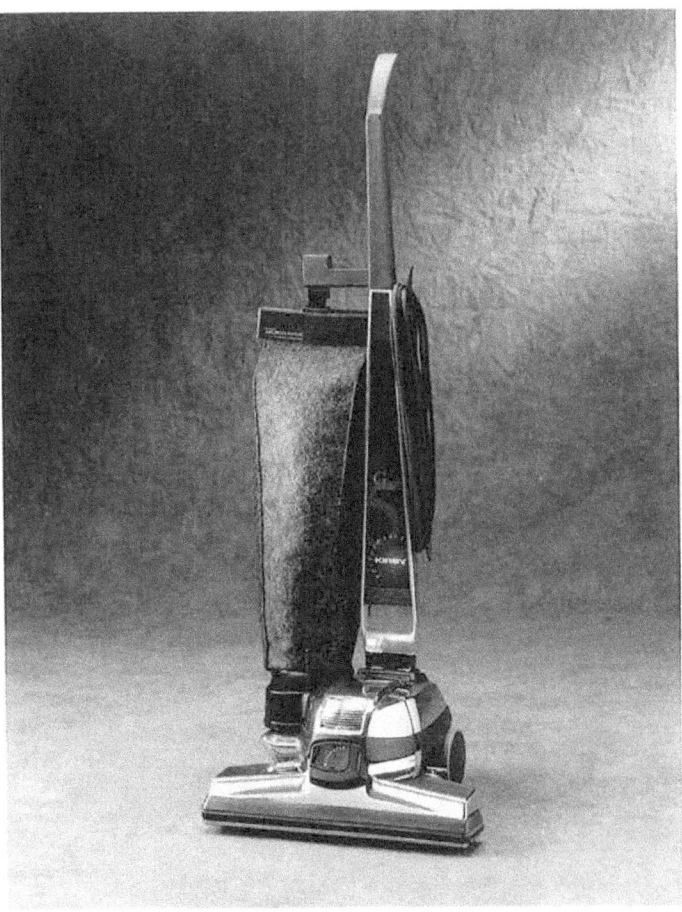

The Kirby G5 incorporates design enhancements, such as a new fan blade, made in cooperation with Lewis Research Center.

Tankless Water Heater

The need to cure water heater woes brought together Kennedy Space Center specialists with Space, Energy, Time Savings (SETS) Systems, Inc. of Miami, Florida.

SETS Systems had designed and developed an electronic "tankless" water heater. Although tankless water heaters have been in use in Europe and Asia for many years, these systems are not efficient enough to run an entire home.

SETS Systems merged computer chip technology with the water heater. By electronically controlling the heating of the water through heater elements, hot water is available when and where you want it—on demand. No more turn-the-knob battles with other water users in the house.

The SETS heater unit was built to efficiently serve the entire home even while several showers and faucets run simultaneously. The SETS tank is designed to render energy savings of up to 50-percent off hot water heating bills. Measuring a modest 12 x 11 x 2-inches and weighing 9-pounds, the unit is a space saver compared to the common tank type water heaters. To remain competitively priced, the SETS Systems water heater was devised to keep manufacturing cost to a minimum, but retain high quality for the consumer.

But Carlos Cabrera, president of the company, had a puzzle on his hands. The flow switch on his tankless water heater design suffered intermittent problems. Hiring several testing and engineering firms produced only graphs, printouts, and a large expense, but not the hoped-for fix to the problem.

Cabrera then heard about the Kennedy Space Center (KSC)/State of Florida Technology Outreach Program. This NASA network, a part of the NASA Southeast Technology Transfer Alliance, runs throughout Florida to provide technical service to businesses at no cost. The program applies scientific and engineering expertise originally developed for space applications to the Florida business community.

Through a Technology Transfer Agreement (TTA), the KSC Technology Programs and Commercialization Office took on the water heat flow switch concern. Picking up the task was engineer Michael Brooks, a 21-year space program veteran. At KSC, Brooks has worked exclusively on the engineering support contract held by I-NET Inc., in support of KSC's Engineering Development Directorate.

Upon scrutiny of the SETS heater unit and flow switch design provided by Cabrera, Brooks discovered key problems with the switch. The solutions were modest, yet needed the keen eye of Brooks, backed by many years of working with flowmeters.

A prototype heater flow switch, incorporating the fixes, was built. Extensive testing of the new assembly worked flawlessly. Older heaters could be fixed simply by replacing two parts in the flow switch without removing the heater.

The heater itself underwent various tests. The new design turned out simpler, yielding a 63 percent reduction in labor and material costs over the old design.

The SETS Electronic Tankless Water Heater is being marketed throughout the United States and worldwide. Cabrera is hoping to expand the family-owned business beyond a $1 million company. Exports throughout Latin America are one action item on a strategic marketing plan.

NASA's assist in solving SETS Systems water heater worries was immediately applauded by Cabrera. "The value they have brought to my company is immeasurable. If every other company in Florida knew about the benefits that this program offers, they'd be crazy not to do it," he says.

NASA personnel helped SETS Systems solve an engineering problem in its tankless water heater. The system heats water on demand instead of storing it in a tank, and is efficient enough to serve an entire home.

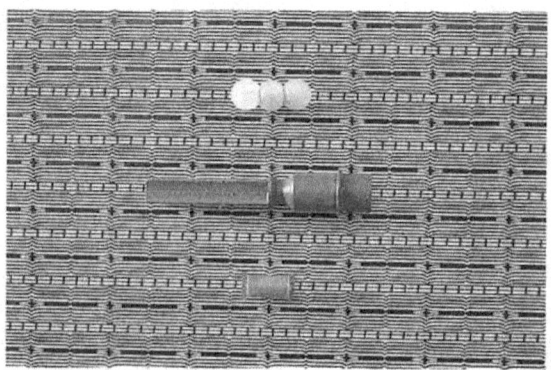

The flow switch developed with NASA expertise (top) solved the intermittent difficulties experienced with the previous designs (bottom). The new switch fits into older systems without other modifications.

Refrigeration Showcases

K eeping supermarket aisle shelves loaded with perishable products colder and floral arrangements fresher calls for well-insulated showcases. Knowledge and technology gained through operating spacecraft in extremes of cold and heat proved invaluable in constructing better performing refrigerator cases.

Through the Technology Affiliates Program at the Jet Propulsion Laboratory (JPL), valuable modifications were made to refrigerator displays built by Displaymor Manufacturing Company, Inc. of Los Angeles, California.

Displaymor, an inner-city manufacturer, could address stiffer federal requirements that ensure the freshness of foods by working with JPL. The application of space technology meant that the small business would be able to continue to market its cases without incurring expenses that could threaten the continued viability of the business, and the future of several dozen inner city jobs.

Displaymor is proud of its cutting-edge refrigerator showcases. By merging integrity, innovation, image, and customer focus, the company's promotional materials assure its customers greater sales with Displaymor products.

Founded in 1960, Displaymor had expanded its product line to cover a broad range of applications, including a full line of floral, egg and all purpose showcases. Using the firm's cases, retailers can cross-merchandise a wide variety of refrigerated products such as ethnic foods, salads, pizzas, dairy and deli-products, yogurt, juices, wine and beer, produce and natural foods. Displaymor also specializes in custom and specific need applications for its clients.

In 1995, new ownership began a reorganization program which included the implementation of a management team and plant modernization. Part of that upgrading involved entering the Technology Affiliate Program with NASA. The results proved profitable for the company.

Research and development improvements in air flow distribution and refrigeration coil technology contributed greatly to certifying Displaymor's showcases given the new federal regulations. These modifications resulted in a refrigerator case that will keep refrigerated foods colder, longer. Such changes proved fruitful, while still maintaining the openness of the display, critical to customer visibility and accessibility, impulse buying, and cross-merchandising.

Displaymor found the association with NASA very productive, strengthening its leadership role in technology development into the 21st century.

Space technology know-how to keep a spacecraft from freezing or frying was applied to refrigerator showcases that ensure that fresh products reach the consumer.

Personalized Learning Software

Students, educators, as well as parents or guardians can find a software assist from products designed by Analysis and Simulation, Inc. (AnSim) of Buffalo, New York.

AnSim's IEPLANNER™ and TPLAN™ products are interactive computer-based systems. They can be run either independently or together as one complete system. Utilized as an Individual Education Plan tool, a user of IEPLANNER and TPLAN can define a goals list, while identifying a host of student demands in motor skills, social skills, life skills, social issues, even legal and leisure needs in the user's area. This computerized, expert tutor and advisor allows assessment of the status of the student and the degree to which his/her needs are being met.

These software tools made use of CLIPS (C Language Integrated Production System), a NASA-developed expert system shell which originated at Johnson Space Center. NASA offers a means of reducing automation costs through a special type of spinoff

service operated by the Computer Software Management and Information Center (COSMIC®). COSMIC supplies to American businesses, like AnSim, at relatively low cost, government-developed computer programs that have secondary utility.

As a company goal, AnSim has focused on the formation of software technology applications for both contractual and commercial products utilizing capabilities in graphical user interfaces (GUIs), expert systems, and simulation. Paul Patti, president of AnSim says the company is exploring new ways to achieve more effective access to the vast and ever increasing amount of information available electronically. Strides by AnSim in computer software have been aided by Johnson Space Center Phase II Small Business Innovation Research (SBIR) funding and NASA work performed in expert systems.

Patti notes that wide area networks, such as the Internet, or just on large local system disks, vast quantities of data can be found. But how best can a

AnSim President Paul Patti watches as Jeffrey Meade demonstrates the graphical user interfaces capabilities of the company's software. AnSim developed the systems using CLIPS, a NASA-developed expert systems shell.

Social workers like M. Daniela Mariano use AnSim's IEPLANNER to assess and identify students' needs in order to develop individualized plans. The company is extending its work through a NASA Small Business Innovation Research contract on a World Wide Web 3D browser.

person sort through huge collections of material, then seek and rank information only of relevance that satisfies a distinctive need? That task can be likened to locating the proverbial "needle in the haystack."

NASA SBIR funding has supported AnSim's Human Memory Extension (HME[tech]) technology. HME[tech] software uses spreading activation models of human memory processes applied to context-sensitive information associations to characterize, rank, retrieve, and recharacterize information. The HME[tech] system automatically and transparently builds and updates a persistent Memory Extension (ME) database of accessed documents. Document titles and other context summaries are parsed, meaning that grammatical form and function, a word or words in a sentence, are maintained by the HME[tech]'s global database.

A user of this indexed meta-database can then retrieve ranked lists of documents via a key word or term. A relevance feedback stage in the software permits the user to quickly zero in on pertinent documents.

A NASA Phase I SBIR has extended AnSim's work in the area of creating a World Wide Web 3D browser. This effort will develop a software prototype of a 3D interactive information visualization system. The software will integrate the ME technology for doing information ranking with a 3D windows-based interactive graphic user interface for doing information visualization.

[TM] IEPLANNER and TPLAN are trademarks of Analysis and Simulation, Inc.
[®] COSMIC is a registered trademark of the National Aeronautics and Space Administration.

Safeguarding Porpoises

To help protect and preserve the harbor porpoise, a low-cost, easy-to-use acoustical pinger stands to gain wide acceptance in the fishing industry. For decades, harbor porpoises have been killed in fisheries, such as those located in the Gulf of Maine. Each year, porpoises have fallen victim to sink gill nets, static fishing devices by commercial fisheries that are meant to catch bottom-dwelling fish in near-shore waters.

Taking on the challenge of lessening the incidental catch of harbor porpoises, the Dukane Corporation's Seacom Division, based in St. Charles, Illinois has designed the NetMark™ 1000.

The NetMark 1000 employs technology originally developed in the late 1960s by NASA engineers at the Langley Research Center. At that time, an underwater location aid was crafted, able to withstand high impact, then emit multidirectional signals for hours on end. Its key purpose was for use in the retrieval of NASA payloads following watery touchdowns on Earth.

Dukane Corporation and Burnett Electronics of San Diego, California later obtained a license from NASA, further improving on the beacon design. Dukane has sold well over 100,000 units since.

A variety of applications for Dukane pingers have included: attachment to "black box" flight recorders on commercial airliners, marking underwater sites and relics and helping to ensure recovery of hazardous cargo in case of accidental loss.

Thanks to a team from Dukane, an underwater acoustic pinger is finding new duty in safeguarding the harbor porpoise. At stake was finding a balance between regulatory rulings, including the Marine Mammal Protection Act, as well as the livelihood of a New England fishing community.

A large-scale field experiment off the coast of New Hampshire in autumn 1994 gauged the effectiveness of acoustic pingers in reducing incidental mortality of harbor porpoises in sink gill nets. The study was carried out by the New England Aquarium, along with researchers from Woods Hole Oceanographic Institution, the New Hampshire Commercial Fishermens Association and the University of New Hampshire's Department of Ocean Engineering.

Dukane provided the specially equipped acoustic alarms for the two-month-long experiment. Broadcasting a signal well within the hearing range of harbor

Originally developed by NASA engineers as an underwater location aid, the fishing industry uses pinger technology to safeguard porpoises from net entanglement. Dukane Corporation's NetMark 1000 emits signals that warn the mammals of the net's presence.

porpoises, the pingers provided warning as to position of the sink gill nets, alerting the mammals to help them avoid net entanglement.

The study found that the use of acoustic alarms appears to hold considerable promise in reducing the number of harbor porpoises killed in sink gill nets in the Gulf of Maine. Indeed, a "dramatic" reduction in the entanglement rate of harbor porpoises in the area was reported.

Following the promising results, Dukane has put in place a manufacturing program to produce the low-cost NetMark 1000 pinger. The sausage-shaped hardware—measuring 6.5 inches in length and 2.3 inches in diameter—has been fabricated to take the beating of deck handling and deployment. Capable of projecting a signal every four seconds over a 100-meter radius, the NetMark 1000 operates down to 100 fathoms, powered by four replaceable standard AA Alkaline batteries. Upwards of 35 days of continuous use can be reached before battery change.

In January of 1997, the NetMark 1000 was selected by the editors of *National Fisherman* for the magazine's "Best of Technology/1996" list, which highlights a collection of the best new gear presented to the fleets during the previous 12 months. Dukane views the NetMark 1000 pinger as the potential standard for an effective, inexpensive acoustic device for reducing porpoise bycatch. Applications of the pinger are not only expanding internationally but to other animal species as well.

™ NetMark is a trademark of the Dukane Corporation.

John Williamson, manager and regulator for the New England Fisheries Council, attaches the NetMark 1000 to a net in preparation for an in situ test. NASA technology is helping the Council find a balance between regulatory rulings and the livelihood of the local fishing community.

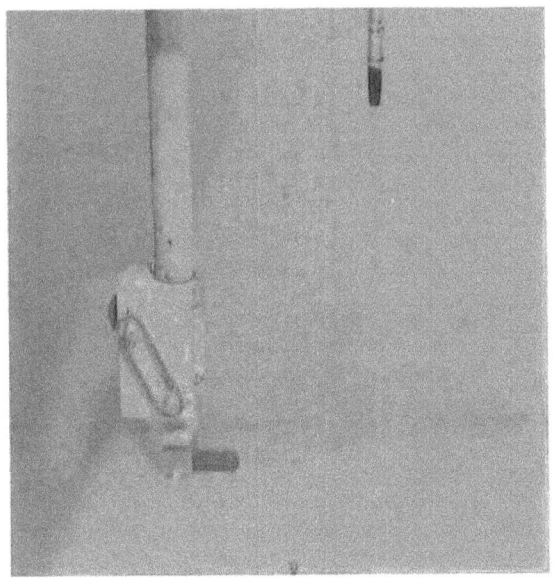

During a test at the University of New Hampshire Ocean Engineering Laboratory, the pinger is dropped to 10 feet below the surface of the water. Scientists then measure the sound pressure waves and frequencies at varying distances.

Automated Pollution Control

Managing a spacecraft built to probe the planet Saturn and its entourage of moons has resulted in the world's first electronic trading mechanism that helps reduce pollution problems, the Automated Credit Exchange (ACE).

The Jet Propulsion Laboratory's (JPL) Cassini mission to Saturn is outfitted with a dozen different research instruments to orbit the planet and drop a probe onto Saturn's moon, Titan. An early challenge of the project was allocating the spacecraft's resources—funding in several fiscal years, different data transmission rates, mass, and multiple power modes—among the instruments and probe.

To support group decision-making, a system was developed named the Cassini Resource Exchange (CRE). The academic design and testing of the CRE was managed largely by David Porter of the California Institute of Technology (Caltech).

The rationale behind the CRE stems from a need to ration spacecraft resources. For instance, scientists who need more power to operate their instruments may trade mass or computer time with another investigator for the additional power.

Typically, a science instrument manager does not have the information needed to efficiently allocate the various spacecraft resources to all the instrument teams.

Since the information necessary for that evaluation is known only by the instrument team, the teams report their resource requirements to the spacecraft manager. Because teams' reports must be conservative in their initial estimates to protect themselves from unforseen issues, teams have an incentive to request resources that may exceed what their Cassini instrument needs.

Therefore, if the science instrument manager were simply to ask the instrument teams how many resources they needed, the messages received from them would lead to an excess allocation of resources.

To counter this situation, the CRE was created. It removes the need for the science instrument manager to know the individual instruments' requirements for the spacecraft resources. Instead, by utilizing several principles of exchange, the CRE induces the instrument teams to reveal their requirements. In doing so, they arrive at an efficient allocation of spacecraft resources by trading among themselves. In essence, CRE promoted bartered exchange between Cassini instrument teams to arrive at "fair" market prices.

NASA funding permitted the CRE to operate through Caltech's Division of Humanities and Social Sciences. The Cassini CRE resides on the Internet, with every instrument scientist having a trading account. Caltech has licensed the CRE methodology and algorithms for other uses.

Patterned after the CRE, Sholtz & Associates of Pasadena, California established an Internet-based concept that automates the auctioning of "pollution credits" in Southern California. The Automated Credit Exchange, or ACE, was launched in April 1995.

ACE began by trading South Coast Air Quality Management District credits for sulphur oxides and nitrogen oxides. A Southern California based RECLAIM air pollution credit trading market was set up. For every pound of a certain pollutant, producers must have one credit. The credits put a cap on the total amount of pollution that can be legally produced in the area, with that amount becoming more stringent each year through 2003. If a company is effective at reducing emissions, its credits can be sold to others who may not be so successful.

Anne Sholtz, ACE's founder, reports that the pollution reduction strategy is proving successful. "Companies are faced with having to make very important environmental control and capital expenditure decisions involving millions of dollars. As a result, they have to balance the cost of reducing emissions by installing new technology with purchasing pollution credits from other companies that have made additional emission reductions," she says.

"The increased number of companies participating in the ACE market is solid evidence that the RECLAIM concept is working," Sholtz adds.

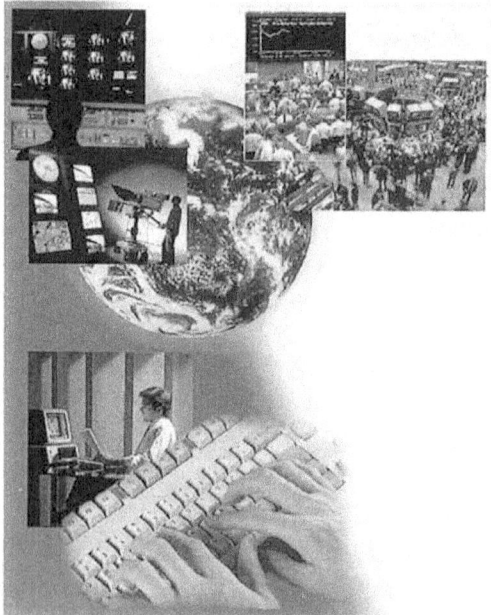

Utilizing space resource allocation technology, the Automated Credit Exchange system helps companies keep pollution and costs down.

Contaminant Monitor Laser

Earth's delicate and protective biosphere can be surveyed using precisely tuned lasers that can detect various chemical compounds and atmospheric pollutants.

Supported by a Langley Research Center Small Business Innovation Research (SBIR) contract, OPOTEK, Inc. of Carlsbad, California developed a laser transmitter for remote sensing of water vapor in the upper atmosphere. The major challenge of the NASA SBIR contract was to develop a solid state laser transmitter which is tuned to a specific wavelength with an extremely narrow spectral linewidth.

A transmitter that can generate very short pulses of continuously tunable laser radiation can be applied to the study of particulate matter within the atmosphere.

Differential Absorption Lidar (DIAL) is a remote sensing technique in which two beams are sent towards the target area. The wavelength of one beam is tuned to match the absorption of the target compound. The second beam is tuned to "miss" the absorption line. The signals are scattered back by the DIAL system by the particles (aerosols) in the atmosphere, and they are detected and analyzed at different arrival times. The ratio of the amplitude of the two return signals, together with the arrival time, provides quantitative information regarding the concentration and the location of the interrogated compound within the atmosphere.

As a leader in developing and using DIAL systems to monitor ozone and water vapor in the atmosphere, NASA was interested in upgrading the capabilities of its airborne laser systems.

At the heart of OPOTEK's work with NASA was an Optical Parametric Oscillator (OPO) that converts a fixed wavelength laser into a tunable source. The output of the OPOTEK MagicPrism™ OPO, which consists mainly of non-linear crystals in a ring cavity, can be tuned continuously over a wide spectral range. This can be done by changing the angle between the optical axis of the crystal and the direction of the beam.

An OPO-based system could replace NASA's current use of dye-laser systems. For OPOTEK to meet NASA's requirements, the OPO was seeded by a narrow-line diode laser, tuned to the absorption lines of water vapor.

OPOTEK is exploring the application of its NASA work to industrial settings. Laser Induced Fluorescence (LIF), for instance, is a powerful technique to monitor chemical processes in extreme environments. As a direct spinoff from its NASA-supported research, OPOTEK began eyeing this technique for industrial applications. In the LIF approach, the laser is tuned to a specific absorption line of the compound which needs to be monitored. Quantitative information is collected by observing fluorescent light emitted from the chemical as a result of the irradiation by the laser beam.

The laser transmitter developed for NASA was used for measuring water vapor in the infrared region. By broadening this concept to other wavelengths, OPOTEK believes a range of industrial applications can be met. A host of other government uses for the technology are also being examined as follow-up by the company.

OPOTEK's tunable laser systems can be used by the Drug Enforcement Administration for discerning the by-products from illegal drug manufacturing. OPO technology has a wide range of applications for Department of Defense in the post cold war era, particularly for local conflicts and terrorism.

Lastly, lasers are in great demand by universities and industrial laboratories. By tuning the laser to interact with specific elements or chemical compounds, it can be used for basic research in photo-chemistry, photo-biology, as well as for diagnostics and monitoring various chemical processes. OPOTEK estimates that the market for scientific lasers alone is on the order of $150 million, of which approximately $70 million are tunable lasers.

™ MagicPrism is a trademark of OPOTEK, Inc.

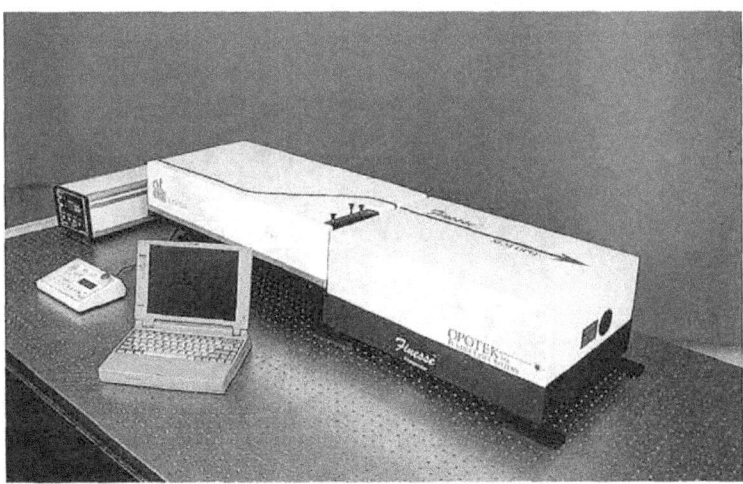

Tunable laser systems built by OPOTEK can monitor Earth's atmosphere for contaminants.

Tire Recycling

The phrase, "where the rubber hits the road" has taken on new meaning for Cryopolymers, Inc. of St. Francisville, Louisiana, near Baton Rouge. They tapped NASA expertise to improve a process for recycling vehicle tires, to convert shredded rubber into products that can be used in asphalt road beds, new tires, hoses, and other products.

The work was done in conjunction with the Southern Technology Applications Center (NASA's Southeast Regional Technology Transfer Center) and John C. Stennis Space Center in Mississippi. NASA expertise in cryogenic fuel-handling needed for launch vehicle and spacecraft operations was called upon to improve the recycling concept.

Stennis advised Cryopolymers on types of equipment required, as well as steps to reduce the amount of liquid nitrogen used in the process. They also guided the company to use more efficient ways to control system hardware.

The process put in place by Cryopolymers utilizes liquid nitrogen to freeze tire scraps to super-cold temperatures—to a chilling 200 degrees below Fahrenheit. This procedure separates reinforcing steel belts and polyester fibers from the rubber. What is left from the treatment is called "crumb," a material that can be further divided into various grades depending on particle size. The larger particle crumb can be used as a component to improve the wearability of a road surface. Also, large particle crumb can be reprocessed to mold products that are low in strength, but high in weather-proofing value.

Even the smaller pieces of crumb are useful. They can contribute to new tires, agricultural hoses, or when mixed with plastics, culvert piping and protective mats for the bed of pick-up trucks.

For each pound of rubber salvaged, nearly 90 percent is reduced into crumb. What remains is called "fluff," scrap metal and polyester residue that can be incorporated into new products as a reinforcing fiber.

"NASA's assistance to Cryopolymers demonstrates the value and wisdom of leveraging federally-funded technology and know-how for private sector use," points

Recycled tire "crumbs" help reduce tire disposal problems and are used in asphalt roadbeds and other items. NASA experts on cryogenic equipment made the recycling idea more cost effective.

NASA know-how in handling super-cold fluids is allowing Cryopolymers, Inc. to recycle shredded tires more efficiently into "crumb." The material can be used to manufacture agricultural hoses.

out Kirk Sharp, NASA Technology Transfer Officer at Stennis. "As in this case, often the bottom line amounts to jobs, increased profits and enhanced global economic competitiveness for the U.S. companies."

"We are so proud of what we are doing," says Joe Kelley, director of community affairs for Cryopolymers.

In terms of value, the Cryopolymers recycling process has proven itself a winner on several fronts. It is estimated that, on a nationwide basis, more than 300 million tires per year are produced. Tire recycling is highly desirable and of great benefit in today's environmentally conscious society.

Cryopolymers expects to reach a production rate of 5,000 pounds of rubber per hour. That translates into over 5,000 tires recycled per day. The new company has created dozens of new jobs in the area and anticipates an annual income of $4 million.

When mixed with plastics, another use of the material generated from recycled tires is bed liners for trucks.

Voltage Controller

Most alternating current induction motors squander energy. Electric motor drive systems are estimated to consume over half of all electricity in the United States and over 70 percent of all electricity in industrial applications. And to make a bad situation worse, energy costs are climbing. Enter Power Efficiency Corporation's soft start energy saving motor controllers.

Based in Hackensack, New Jersey, Power Efficiency Corporation was specifically formed to manufacture and market products exclusively developed from NASA technology.

The original idea for a three-phase power factor controller with induced electric and magnetic fields sensing was developed by Frank Nola, an engineer at Marshall Space Flight Center. Patented by NASA in 1984, Power Efficiency Corporation later licensed the technology.

Power Efficiency and two major distributors, Performance Control located in Ann Arbor, Michigan and Edison Power Technologies of Paramus, New Jersey, use the electronic control boards that represent the NASA developed technology to assemble three different motor controllers.

These motor controllers are known by brand names: Power Commander, Performance Controller^{SM}, and Energy Master. All three products have been accepted by the marketplace. Customer lists include a large number of multi-billion dollar companies, such as May Department Stores, Caesers Atlantic City, Ford Motors, and American Axle.

Power Efficiency's Power Factor Controller (PFC) reduces excessive energy waste in AC induction motors. The motor controller is a solid state unit providing a reduced voltage start, reduced energy consumption and improved power factor. The unit monitors the phase lag of the current and voltage relationship in a motor that is operating at less than full mechanical load. The controller cuts back the voltage to precisely what the motor requires to maintain the rated speed and torque under the present load.

Volts, amps and watts are reduced and motor life is increased. The results show up immediately as a financial savings at the utility meter. The significance of the PFC lies in the fact that nearly a billion induction motors are used daily. The motor controller is used in industries and applications where motors operate under variable loads, including elevators and escalators, machine tools, intake and exhaust fans, oil wells, conveyors, pumps, die casting, and compressors.

In January 1997, Power Efficiency filed a patent application with the U.S. Patent and Trademark Office for its newly developed phase detector technology.

Motor controller to reduce energy consumption in AC induction motors was developed by NASA. The technology has been licensed and is being widely used in many commercial applications.

"The reason for the change is based solely on the engineering changes that have occurred in the manufacture of AC Induction motors," says Power Efficiency Corporation president, Nicholas Anderson. "The original NASA technology was excellent for the control of AC Induction motors manufactured in prior years. The new patent applied for serves the specific purpose of controlling motors that were not in existence when Frank Nola received his patents, and the patent we have applied for would not exist without the NASA patents and technology," Anderson says.

Recently, Power Efficiency Corporation became a public corporation and is traded on the NASDAQ. While the offering was small, but successful, the company is poised for future growth.

"We are truly indebted to NASA for our start and our success," says Anderson.

^{SM} Performance Controller is a salesmark of Power Efficiency Corporation.

Infrared Fiber Optic Sensors

Remote sampling of hazardous environments or judging the quality of paper, textiles, and composites on a production line are among the duties of an innovative infrared fiber optic probe.

The creation of remote infrared fiber optic sensing systems by Sensiv, Inc. of Waltham, Massachusetts was assisted by successive years of Small Business Innovation Research (SBIR) funding from Langley Research Center. Sensiv is a joint venture between Foster-Miller, Inc. of Waltham and Isorad, Ltd. of Israel.

NASA's SBIR interest in infrared, fiber optic sensor technology was geared to monitoring the curing cycles of advanced composite materials. Foster-Miller was awarded Phase I and Phase II SBIR contracts to develop the In-Situ Fiber Optic Polymer Reaction Monitor that could lead to higher yields and lower costs in complex composite manufacturing. These funds helped in the fabrication of an infrared, fiber optic sensor to track the

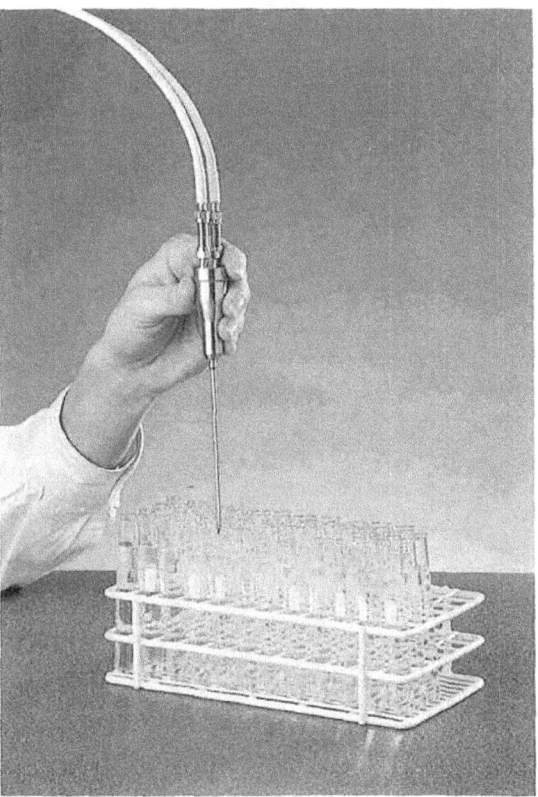

Fiber optic sensing system by Sensiv has made chemical, environmental and biological sampling in remote and difficult areas possible. The NASA spinoff is rooted in space research that studied how best to monitor the curing of composite materials.

molecular vibrational characteristics of a composite part while it is being cured.

The inventive sensor utilizes Fourier Transform Infrared (FTIR) spectroscopy. This is a widely accepted laboratory technique for performing precise compositional analysis in solids, liquids, and gases. A major drawback, however, was the need for expensive sample holding accessories and labor intensive sample preparation. These barriers confined use of the method to the laboratory.

Placing this highly accurate FTIR analytical tool directly at a processing site or production line is central to Sensiv's remote infrared fiber optic sensing system. This allows more rapid study of defects and field samples, eliminating the time and cost of returning samples to a central laboratory.

Foster-Miller ingenuity allowed infrared transmitting optical fibers to combine with FTIR spectroscopy to enable remote sensing. How it works is straightforward. The optical fiber gathers data via reflected light, which is transmitted to the spectrometer through a cable. Flexible fiber optic cables permit the probe to reach difficult to get at samples. The spectrometer may be over 15 feet away from the processing line or sample. Measurements can be taken every few seconds depending on the type of process being monitored. Spectra are gathered by computer for analysis.

Sensiv probes operate in the mid-infrared range of the spectrum, although modifications to the instrument also permits its use in the near-infrared region. The Sensiv needle-probe is built to be placed in a liquid or powder and analyze the chemicals in the mixture. The small size of a needle-probe, for example, is ideal to get samples such as circuit board components, recessed sample areas, skin, small sample containers and large samples such as metal parts and forensic samples.

The pencil grip-like design of the needle-probe permits the acquisition of excellent quality spectra from samples in many different orientations.

Probes, cables and spectrometers used in the system can be customized to meet specific process needs of a customer. This remote, real-time analytic capacity makes the probe useful for chemical and environmental purposes, as well as in medical applications.

Other applications of the probe system include food processing control; combustion control in furnaces; maintenance problem solving; and numerous defense industry applications such as cure monitoring of solid rocket motor propellants.

Sensiv is working with Spectra-Tech Inc. of Shelton, Connecticut to commercially distribute the FTIR probe system in the United States and overseas.

Monitoring Earth's Atmosphere

E arth's atmosphere is a distinctive blend of chemistry that sustains life here on the planet. With the global population nearing 6 billion, it is believed that human activities are inducing atmospheric change. Because of this linkage, modifications in the atmosphere may be early harbingers of global climate change. To better monitor the atmosphere, from the ground, in the air, and from space, various tools are being employed.

One such tool is a new Micro Pulse Lidar (MPL) System, originally developed by Goddard Space Flight Center and now available commercially from Science & Engineering Services, Inc. (SESI) of Burtonsville, Maryland through a technology transfer license.

Lidar stands for Light Detection and Ranging, an optical analog to radar. Using a lidar, short, intense pulses of laser light are directed to disperse through the atmosphere. As this laser light propagates, it interacts with aerosols—particles of liquid or solid dispersed as a suspension in gas—and molecular constituents. The backscattered energy is measured by the Lidar system for detailed study.

SESI's Micro Pulse Lidar System is built to characterize numerous details of the composition and dynamics of the atmosphere, such as atmospheric cloud and aerosol concentration. The system is suitable for environmental monitoring studies that require full-time, unattended measurements of cloud and aerosol height structure.

A main feature of SESI's MPL System is beam expansion, making it eye safe at all ranges, including the exit aperture of the instrument. Being eye safe eliminates risk to civilians or aircraft, even in continuous operation. The device has impressive sensitivity in signal detection over a great range. The MPL is capable of detecting all significant cloud and aerosol scattering from the ground through the troposphere, and into the stratosphere—a distance of over 15 miles. Designed for long-term unattended operation, the SESI MPL System is well suited for applications requiring routine monitoring of the atmosphere.

Backscattered radiation received by the MPL is transformed into electrical signals which are subsequently converted into digitized data. Data are then collected, stored, and analyzed by the system's rackmounted IBM-compatible personal computer. The MPL PC also coordinates all of the system operations.

Another feature of the SESI Lidar is the instrument's modest dimensions, allowing it to be used where space is limited. The system can also be equipped with a protective climate-controlled enclosure, permitting placement of the system in field operations where adequate sheltering facilities do not exist. Upgrading of the unit can be easily done for greater resolution, performance, or to make the instrument more rugged for use in an aircraft.

Studies of climate dynamics, meteorological research, and environmental monitoring are but a few possible applications using SESI's Micro Pulse Lidar.

SESI has been performing Small Business Innovation Research work for Goddard Space Flight Center and Wallops Flight Facility, as well as private industries. SESI is engaged in the development of unique tunable solid state laser systems, various lidar systems for atmospheric measurements, detection and ranging equipment, and medical instrumentation.

Science and Engineering Services, Inc.'s Micro Pulse Lidar System resulted from NASA research and patents. This instrument is employed to make detailed measurements of atmospheric constituents and is available on the commercial market.

Monitoring Earth's Ecosystems

G lobal satellite monitoring of Earth's biosphere, along with its rich tapestry of land, oceans, and ice has become increasingly important. Our very lives may well depend on better knowledge of Earth's diverse and geographically distinct set of ecosystems.

Partnered with Goddard Space Flight Center, Sensit Technologies Inc. of Portland, North Dakota developed a third-generation Portable Apparatus for Rapid Acquisitions of Bidirectional Observations of Land and Atmosphere, or PARABOLA III for short.

PARABOLA III, now commercially available, is designed to measure the reflected signature of a variety of Earth surface types, from rangeland vegetation to ice and snow. It can rapidly acquire data for almost the complete sky and ground-looking hemispheres, with no missing data or "dead cone" and sufficient dynamic range to measure direct solar radiance.

A unique field instrument, the PARABOLA III, is easily transportable to remote sites. Battery-powered, the apparatus operates in eight spectral bands, taking just four minutes to complete one rotating scan, including the sending of data automatically. It can also scan continuously without interruption.

PARABOLA III was actively used in the Boreal Ecosystem-Atmosphere Study (BOREAS). This intensive one-month field campaign in 1996 concentrated on understanding energy-water-carbon exchanges between the boreal forest and the atmosphere. That campaign involved some 120 scientists and five research aircraft. Study areas were near Prince Albert, Saskatchewan and 400 miles away to the north-east, in Thompson, Manitoba.

Data gleaned by PARABOLA III proved useful in appreciating how the land's vegetated surface couples with the Earth's lower atmosphere, can influence weather in the short term, and climate change in the long term.

PARABOLA III cataloged the multidirectional interactions of solar energy in various types of boreal forest canopies. Through intensive measurements and modeling, instrument data was matched with ecologically important biophysical parameters. This information is proving useful in designing a Multi-angle Imaging SpectroRadiometer (MISR), a satellite instrument that will measure sunlight reflected by the Earth into space. MISR is being built by the Jet Propulsion Laboratory as part of NASA's Mission to Planet Earth program.

Another significant contribution of PARABOLA III, remarks Paul Stockton, President of Sensit Technologies, is that it maximizes the usefulness of "off-nadir" viewing data to be gathered by satellite sensors. Off-nadir means observing objects hundreds of miles off a ground track, typically by rotating a mirror so sensors can look sideways.

In general, large viewing angles provide enhanced sensitivity to atmospheric aerosol effects and to cloud reflectance effects. Appreciating the interchange of radiation to and from clouds, the type of cloud, as well as land surface category is valuable data, allowing more accurate estimates of global climate models. But off-nadir viewing also complicates the satellite analysis of vegetation changes.

PARABOLA III is aiding in the design and calibration of MISR, and other off-nadir satellite sensors being built.

Spherical scanning radiometer built by Sensit is helping Earth remote sensing researchers determine what the atmosphere does to satellite images taken from orbit.

Structural Analysis and Design Software

One-of-a-kind Langley Research Center computer code for designing exotic hypersonic aircraft was transferred to a private company for more pedestrian use in ground transportation, building construction and marine industries.

Working through the Technology Applications Group at the NASA center, the Collier Research and Development (R&D) Corporation of Hampton, Virginia received the first ever Langley software copyright license agreement. The agreement was signed in May 1996.

Collier R&D transformed the NASA computer code into a commercial software package called HyperSizer™. The commercial software package integrates with other popular Finite Element Modeling (FEM) and Finite Element Analysis (FEA) private-sector structural analysis and design packages.

The Langley and Collier R&D agreement is viewed as a pioneering step for government transfer of technology to U.S. industry. Collier R&D will pay NASA royalties from software sales.

"For NASA, it represents the emerging recognition of the value of computer software as a potentially licensable technology. The software intellectual property rights were treated similarly to hardware patent rights," says Collier R&D's Ivonne Collier, president of the company.

"For Collier R&D, the agreement represents a broadening of its business from engineering consulting to developers and marketers of software technology," says the company president.

The NASA software, called ST-SIZE, was chiefly conceived as a means to improve and speed the structural design of a future aerospace plane for Langley's Hypersonic Vehicles Office. Different classes of materials under consideration for use on a hypersonic plane could be computer modeled, then shown how they would react under extreme temperature changes, speeds, pressures and other operating conditions. The software tool gave structural engineers the confidence to select the proper lightweight materials for use in high-speed aircraft.

Including the NASA computer code into the HyperSizer software package has equipped Collier R&D to look beyond aerospace to other high-tech applications. These include improved design and construction for offices, marine structures, cargo containers, commercial and military aircraft, rail cars and a host of everyday consumer products.

HyperSizer can evaluate and optimize:

- any cross sectional shapes, sizes, thicknesses, materials selections, and material layups;
- many composite material types such as polymer, ceramic, metal matrix, as well as concrete, wood, steel, and aluminum alloys;
- thermal stress problems caused by thermal gradients from aerodynamic heating and/or cryogenic fuels; and
- weight estimations and structural integrity. Failure mode checks performed with HyperSizer can recognize potential structural deficiencies of any component early in the project's design phase.

Previously an engineering consulting organization, the addition of HyperSizer has enhanced the Collier R&D Corporation portfolio of services and products, while strengthening its competitive posture within the software industry.

™ HyperSizer is a trademark of Collier Research and Development Corporation.

Collier R&D's HyperSizer software, developed from NASA technology, displays an aircraft's surface in multicolored pixels.

Smart Test Machines

Be it the whine of a race car or the roar from a space shuttle liftoff, also at work is the meshing of materials running at high-speed, under intense pressure and high heat loads. By using lubricants, the inner-workings of such engines can be protected from excessive wear and tear.

For years, Vern Wedeven had been steadfast in his pursuit of "smart" test machines, computer-aided equipment that could mimic the tortuous conditions that bearing and gear hardware undergo. Making the task even more difficult was developing ways to simulate high stress contacts between materials, not on a macro-scale, but down to microscopic level.

Wedeven formerly worked for NASA Lewis Research Center where he carried out studies in bearing lubrication for aerospace products. During his NASA years from the 1970s into the early 1980s, the materials engineer initiated an "Interdisciplinary Collaboration in Tribology" (ICT). Tribology is the science of lubrication and friction between material surfaces in motion.

The effort involved NASA, some six universities and several university professors. Central to the ICT program was crossing the discipline barriers between chemistry, material science, surface physics, engineering design and application. Wedeven's ICT initiative, in fact, linked the tribology disciplines of chemistry, mechanics and materials parts for application to advanced aeropropulsion systems.

This NASA-sponsored ICT work provided the foundation for Wedeven in 1983 to form his own company, Wedeven Associates, Inc., of Edgmont, Pennsylvania. The small firm is now blazing new trails to provide industry with a quick-response problem solving ability and a more proficient means to get tribology into the marketplace.

The company's first smart machine was called WAM1, for Wedeven Associates Machine, version one. While WAM1 would be judged as crude by today's standards, it proved indispensable in characterizing oil quality, matching that value with performance levels in lubricating bearings and gears.

A later variant was the WAM3. This smart machine was tapped by NASA to help increase the life of turbo pump bearings in the space shuttle's main engine. Those bearings were failing while liquid oxygen, maintained at -293 degrees, was being pumped through the engine at speeds comparable to emptying an average swimming pool in just four minutes. Utilizing the WAM3 machine, lubrication improvements and ways to enhance the durability of the shuttle's turbopump bearings have been realized.

Given the rapid growth in computer software and hardware, Wedeven Associates is hard at work on a WAM4 smart machine. This computer-controlled device can provide detailed glimpses at gear and bearing points of contact. WAM4 can yield a three-dimensional view of machinery as an operator adds "what-if" thermal and lubricant conditions, contact stress, and surface motions. The company expects to be issued a patent for both the machine and test method by summer of 1997.

Along with NASA, numbers of firms, including Pratt & Whitney, Caterpillar Tractor, Exxon, and Chevron have approached Wedeven for help on resolving lubrication problems.

While Wedeven and his associates have been building their smart test machines, other benefits have evolved. The company discovered the concept of vapor/condensation lubrication, for instance. This type of lubrication would permit a bearing to function in a severe environment with only a small quantity of liquid lubrication. New methods to screen oils for fuel efficiency and useful life are also being established. Lubricants that work over a wide temperature range have been developed.

This research could well spark a new generation of commercially-available, environmentally friendly, energy efficient engines of the future.

Wedeven Associates developed the WAM4, a computer-aided "smart" test machine for simulating stress on equipment, based on the company president's bearing lubrication expertise gained while working for NASA.

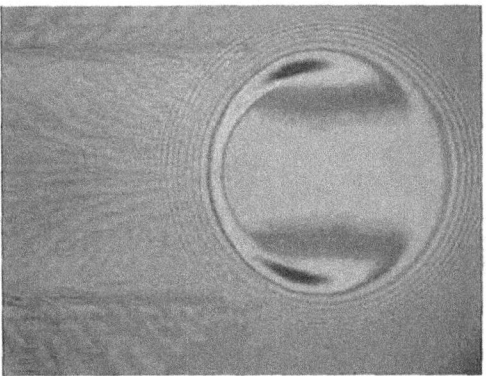

The WAM4 shows the pattern of interference fringes of contact in oil as pressure is applied during a test. Products tested using the WAM4 include Pratt-Whitney jet engine cages and bearings, and gears used in NASCAR racing.

Real-time Simulation

Virtual worlds of stunning realism is the pledge from products created by real-time 3D software manufacturer Coryphaeus Software, Inc. of Los Gatos, California.

Coryphaeus Software was founded in 1989 by former NASA electronics engineer, Steve Lakowske. His 10-years in the space arena included work on a vertical motion simulator at Ames Research Center, and other tasks associated with human and machine interaction. One of his assignments at Ames was design, fabrication, and testing of audio, communications, and aural cue systems in a multimillion dollar aviation human factors research facility.

As president of the privately-owned Coryphaeus, Lakowske turned his prior experiences into a successful software firm producing high-end, real-time simulation products for virtual reality applications. Annual sales are now over $6 million.

"My ten years at NASA allowed me to learn much of what I'm still applying today in this business," says Lakowske. "The technology we were developing in the various projects is something that was rarely done in the commercial sector, so there was a considerable amount of technology transfer, at least in terms of the ideas and principles. And I hope that sort of tech transfer continues," he adds.

Designer's Workbench™ was the flagship product of Coryphaeus, a modeling and simulation tool for the development of both static and dynamic 3D databases. Other products soon followed, such as Activation™, EasyT™ and EasyScene™.

Activation was specifically designed for game developers to play and test next-generation 3D games before they commit it to a target platform. With Activation, game publishers can shorten development time and prove the "playability" of the title, thereby maximizing their chances of introducing a smash hit.

EasyT lets users create massive, realistic representations of Earth terrains that can be viewed and traversed in real time. Importing satellite data into EasyT is easily done, to accurately represent elevation data, as well as carry images of Earth. This software allows users to zoom in on models, perusing data at a high level of detail.

EasyScene software controls the actions among interactive objects within a virtual world. This tool creates and plays back behaviors between the virtual objects and the virtual environment.

All Coryphaeus products are designed to promote high productivity for both engineering and

Designed by a former NASA engineer, Coryphaeus software tools provide the ability to create real-time visual simulations in 3D to train jet aircraft pilots.

non-programming users. The result is easy-to-use tools to establish real-time 3D simulations for interactive instrumentation, out-the-window scenes, virtual reality and game authoring.

Customer applications include real-time flight and driving simulation and training systems, terrain modeling and test range simulations, pilot's view re-creations from flight data recorders, and real-time walkthroughs of facilities.

Coryphaeus products are used on Silicon Graphics workstations and supercomputers to simulate real-world performance in synthetic environments. The company is also working to develop its software to operate on other platforms.

A representative sampling of commercial customers for the company's software tools includes aerospace, aviation, architectural and engineering firms, game developers, and the entertainment industry. Boeing, Hughes Training, Lockheed Martin, ITT Automotive, as well as NASA and branches of the U.S. armed forces, are among the list of clients.

™ Designer's workbench, Activation, EasyT and EasyScene are trademarks of Coryphaeus Software, Inc.

Software tools developed by Coryphaeus yield real-time 3D visual simulations ideal for plotting the twists and turns of a highway for planning purposes.

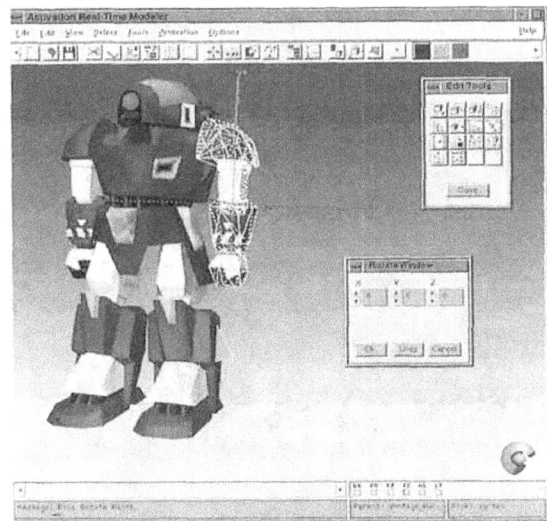

Activation, a real-time graphics software by Coryphaeus, allows a user to incorporate models, animation sequences, and character animation into a game being prototyped.

Miniature Heat Pipes

The widespread and fast-paced adoption of notebook computers in the home, workplace and on-the-road owes part of its success to space shuttle and space station technology.

Thermacore Inc. of Lancaster, Pennsylvania has been associated with Goddard Space Flight Center since 1989. NASA Small Business Innovation Research (SBIR) contracts with Thermacore fostered company work on devices tagged "heat pipes" for space applications.

A heat pipe moves heat from spot-to-spot with little loss in temperature. The heat pipe employs a two-phase process in which a liquid is evaporated inside the pipe by heat input at one end. The resulting vapor is condensed back into the liquid by heat removal at the pipe's other end. The condensate is then returned to the evaporator by capillary action in a wick.

To control the extreme temperature ranges in space, heat pipes are vitally important to spacecraft. Of additional benefit is that heat pipes do not suffer from the wear and tear of moving parts. They are, therefore, exceptionally reliable.

As Thermacore's Chairman, G. Yale Eastman, points out: "All heat pipes are not created equal." Using its own funds, the firm employed knowledge gained from SBIR research to solve a challenge in the commercial marketplace of notebook computers.

The problem was to maintain an 8-watt central processing unit (CPU)—the real brains of a computer—at less than 90 degrees Celsius in a notebook computer using no power, with very little space available and without using forced convection. Using a fan to cool this electronic circuitry required far too much battery power. That would reduce the useful operating time of the notebook.

Thermacore's answer was in the design of a powder metal wick that transfers CPU heat from a tightly confined spot to an area near available air flow. The Thermacore heat pipe technology permits a notebook computer to be operated in any position, even upside down, without loss of performance.

Miniature heat pipe technology has successfully been applied, such as in Pentium® Processor notebook computers. In desktop computers, higher-powered CPU chips or multi-chip modules means higher-powered power supplies, more memory and other power-hungry components. Each of these components dissipates heat. Thermacore expects its heat pipes to accommodate these computers as well.

Given the high volume of computer manufacturing, the cost of heat pipes has been reduced significantly, quite often providing an economical solution to many cooling applications. Thermacore has used company funds to develop the fabrication processes needed and to build the factory required for mass production of miniature heat pipe technology. Production rates have already exceeded 5,000 units per day.

Cellular telephones, camcorders, and other hand-held electronics are foreseeable applications for heat pipes, predicts the company.

By advancing commercial use of federally-sponsored research, Thermacore received in October 1996 the SBIR Technology of the Year Award in the Computer/Electronics category.

® Pentium is a registered trademark of Intel Corporation.

NASA heat pipe technology used in spacecraft to keep hardware and critical electronics cool has found its way into notebook computers. Tiny heat pipes are in wide use to cool the main central processor chip.

Individualized Communications

Innovative ways that a company can reach large audiences on a one-on-one, personalized basis becomes practical using IntelliWeb™ and IntelliPrint™. These products are available from MicroMass Communications™, Inc. of Raleigh, North Carolina.

Established in 1994, MicroMass is a communications software company devoted to helping their clients communicate individually with each member of a company's customer base.

The first commercially-released product from MicroMass is IntelliWeb, a full-featured website personalization tool. MicroMass also provides IntelliPrint, a product designed to create tailored, individualized messages via printed media.

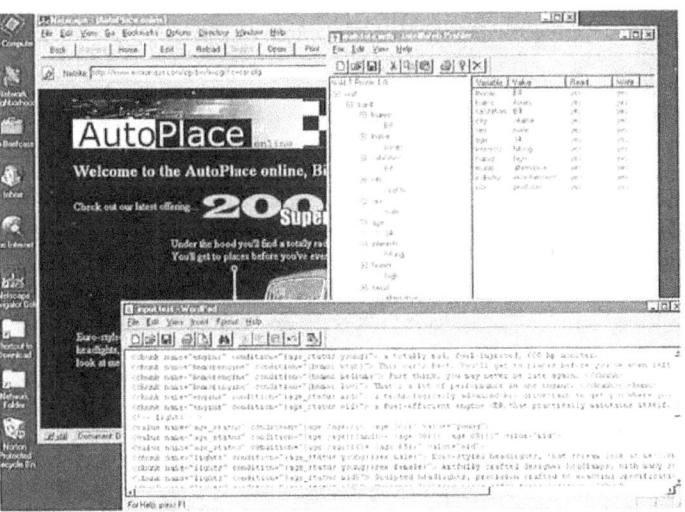

Based on a NASA software system, IntelliWeb allows users to personalize their websites.

At the heart of these MicroMass personalization technologies is CLIPS. Computer technicians at Johnson Space Center produced CLIPS, a development and delivery expert system tool. CLIPS stands for C Language Integrated Production System,. It provides a cohesive software tool for handling a wide variety of knowledge with support for three different programming paradigms: rule-based, object-oriented, and procedural. CLIPS was made available to MicroMass Communications through the Computer Software Management and Information Center (COSMIC®) in Athens, Georgia.

"By embracing CLIPS and its expert system technology, our personalization products can make 'expert' content selections. This has put our products in an elite group," says Mark Rinehart, MicroMass' director of systems engineering.

IntelliWeb delivers personalized messages by dynamically creating single web pages or entire web sites based on information provided by each website visitor. A key feature of the product is that a user need not wade through volumes to get to sentences. IntelliWeb developers tie content databases to expert system-based rules/facts databases. These are activated when a visitor's information is either entered via a standard HTML form or provided from a customer database. Also, the databases are triggered when the user's own personal IntelliWeb profile is supplied.

IntelliPrint uses proprietary technology to generate printed messages that are personally relevant and tailored to meet each individual's needs. Over time, IntelliPrint establishes a dialogue with each customer using individual customer feedback. This information is then utilized to create timely, relevant and enduring personalized messages.

"IntelliPrint dramatically changes the way people have traditionally communicated," says David Bulger, MicroMass chairman and CEO. Historically, people have only been able to communicate on either a micro (one-to-one) or mass (one-to-many) level, he explains.

Examples of IntelliPrint in operation are numerous. For instance, Bristol-Myers Squibb, in conjunction with MicroMass, developed a personalized newsletter, *Living at Your Best*. The newsletter content was geared to each recipient based on a health and lifestyle survey taken earlier.

IntelliPrint also is the core technology behind SmithKline Beecham's Nicorette Committed Quitters[SM] Program. This 12-week program was built by MicroMass using IntelliPrint software, coupled to the latest in demand printing technology. Participants are provided individually customized motivational materials to support their attempt to stop smoking.

"Our system combines the intimacy of very personal, micro communication with the reach of mass communication, enabling our clients to develop and sustain interactive conversations with millions of people," Bulger says.

[TM] IntelliWeb, IntelliPrint and MicroMass Communications are trademarks of MicroMass Communications, Inc.
[®] COSMIC is a registered trademark of the National Aeronautics and Space Administration.
[SM] Nicorette Committed Quitters is a salesmark of SmithKline Beecham.

Managing Satellites

Operating a satellite requires 24 hour-a-day, 7-day-a-week tender loving care. Telemetry must be acquired routinely, and a watchful eye must be ever present to track and control a spacecraft.

EPOCH 2000™ is from the product line of Integral Systems, Inc. of Lanham, Maryland. This computer software allows ground operators to monitor and control satellites over a wide area network. The software decreases the costs of managing satellites once in space by automating such functions as telemetry processing, commanding, anomaly detection, and archiving collected data.

As a third generation product from Integral Systems, EPOCH 2000 owes part of its heritage from work the company completed to support Goddard Space Flight Center. A decade of valuable NASA experience was incorporated into the company's EPOCH 2000 product line for satellite command and control, says Integral Systems vice president of commercial systems, Steven Carchedi.

Perhaps most people think of satellite control centers as large rooms, packed wall-to-wall with rows of distressed technicians and giant number-crunching computers. EPOCH software is run from individual workstations, tied together via a local area computer network. These workstations operate in an office environment, meaning there is no need for an expensive computer room, power service, or cooling. There is no single point of failure as each workstation operates independently. EPOCH software can be run on essentially any host computer and still maintain communications with all the other processing nodes.

The system is designed to operate multiple satellites simultaneously. Since EPOCH 2000 is completely database driven, it can be used for any satellite or ground station configuration.

To meet the NASA challenge, Integral Systems was selected by the Johns Hopkins University Applied Physics Laboratory to support the first NASA Discovery-class mission: the Near Earth Asteroid Rendezvous (NEAR) program. The rendezvous will take place in early 1999 and the NEAR spacecraft will orbit the Asteroid Eros for a year. EPOCH 2000 forms the core of NEAR's command and control ground system back on Earth.

Integral Systems was selected in November 1996 to provide up to 15 Low Earth Orbit Autonomous Ground Terminal (LEO-Ts) for NASA. The LEO-T systems are to be delivered under contract to NASA's Wallops Flight Facility at Wallops Island, Virginia. The LEO-T is designed to make it easier and less expensive for principal investigators to obtain telemetry, tracking and control services for their science missions.

Founded in 1982, Integral Systems has grown to be a leading provider of satellite command and control systems to government and commercial customers, such as AT&T, GE Americom and TRW. Company revenues have climbed to over $11 million, with over 100 employees working at its headquarters. Its commercial-off-the-shelf satellite software products have supported well over 70 satellite missions aimed at scientific research, meteorology, or communications applications.

™ EPOCH 2000 is a trademark of Integral Systems, Inc.

Integral Systems, Inc.'s EPOCH 2000 forms the core of NASA's Near Earth Asteroid Rendezvous mission's command and control ground system. Managed by the Johns Hopkins University Applied Physics Laboratory, the satellite was launched in February 1996.

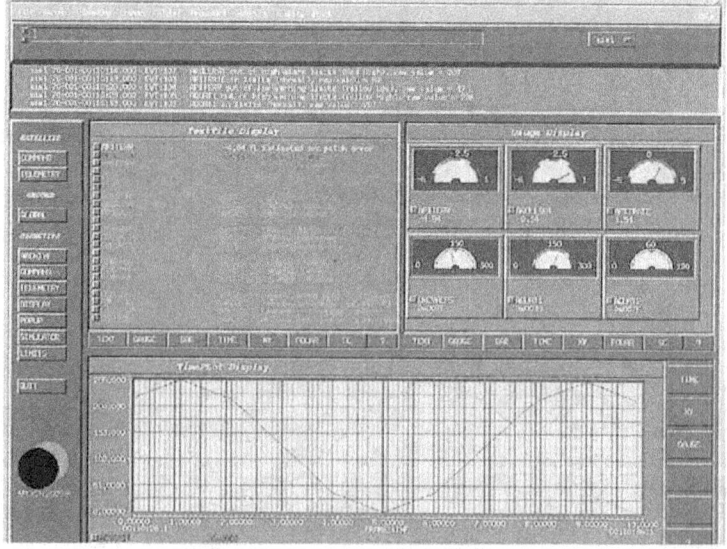

Internet Business Solutions

The dictionary defines the word "cogent" as "convincing, compelling belief." For David Atkinson, president and CEO of Cogent Software, Inc. in Pasadena, California, it is his belief that the Internet is changing the way we do business. It is a personal conviction of Atkinson that you can't let your competition leave you behind in the information revolution.

Cogent Software was founded in January 1995 to support customers with Internet business solutions. Joining Atkinson in founding Cogent is Irene Woerner. Both individuals, as well as several other Cogent employees, formerly worked at the Jet Propulsion Laboratory (JPL). The NASA-developed expertise of Cogent's key personnel, according to Atkinson, is one key to the company's early success in providing low-cost, high-quality service and completed projects.

Atkinson headed JPL's Information Systems Technology section, managing operations technology and artificial intelligence program development for NASA. Woerner led JPL's Advanced User Interfaces Group, where she conceived and directed the research and development of NASA hypermedia library systems.

Cogent Software's mission is to help companies organize and manage their online content by developing advanced software for the next generation of online directories and information catalogs.

While still a young entrepreneurial start-up, Cogent already offers a complete range of Internet solutions for businesses, including Internet access, Web site design, local and wide-area network solutions, and custom software for online commerce applications. Complete services for businesses nationwide include national dial-up access, dedicated high bandwidth connections, and special email services for businesses.

As a service, Cogent is providing DesignSphere Online. "DesignSphere Online aims to help build an electronic community for the communications arts industry," says Atkinson.

From the most seasoned designers to desktop publishers, DesignSphere Online provides access to a variety of creative resources and functions, such as:

- Biz-2-Biz: a searchable business to business directory with free listings that can be enhanced with optional graphics, email links and hyperlinks;
- Showcase: potential clients can preview online portfolios of creative professionals in the areas of illustration, design, photography and new media; and

- What's Hot: summaries of industry's hottest hardware and software technology, with a special feature called Art Mart Stock photography and illustrations from key providers can be browsed online in Art Mart.

Cogent's customers range from small offices to manufacturers with thousands of employees. The group has been selected as the exclusive, endorsed Internet Service Provider by the Electronics Representatives Association (ERA), National Electronics Distributors Association (NEDA), and North American Industrial Representatives Association. Other key customers of Cogent Software now include United Chemi-Con, one of the largest manufacturers of capacitors in the world.

"No one way is right for every business or organization," according to Cogent Software, Inc. Created with expertise gained at NASA, the company's services and products, such as DesignSphere Online, support customers with Internet business solutions.

Data Management System

The Space Shuttle and early Space Station projects provided a major inducement for devising technology and systems integration sources. Huge databases chock full of information from contractors, the tracking numbers of space hardware, and a wealth of related facts and figures must be maintained.

To support these information-handling needs, a product data management, electronic document management and workflow system was designed at Johnson Space Center.

Working with NASA, CENTRA 2000 Inc., a wholly owned subsidiary of Auto-trol Technology in Denver, Colorado, obtained permission to use the original software, helping to move the system into the commercial marketplace. Initially, just 33 database tables comprised the original management software, which was later expanded to about 100 tables.

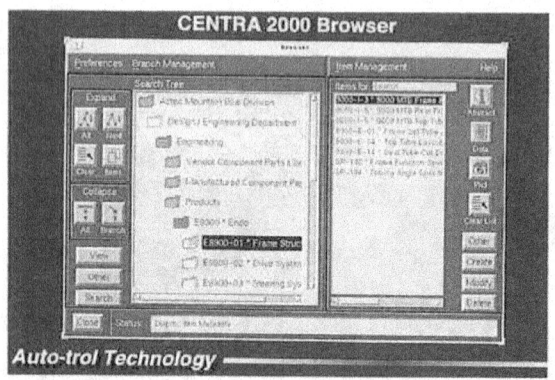

Auto-trol Technology's CENTRA 2000 manages huge amounts of data documentation, computer drawn images, and other project-related work information.

This system is called CENTRA 2000™. "Linking your business to your information" is Auto-trol Technology's slogan that underscores the proficiency of CENTRA 2000.

CENTRA 2000 offers the ability to start simple and then grow as an enterprise grows. Designed for quick implementation, the data management system uses menu driven setup functions. Organizational structures are easily established and initial data can be batch loaded. Many users of CENTRA 2000 are in production within 90 days of installation. That speedy process demands fewer people, less time and less money. Pressures are indeed strong today to bring products to market more quickly, with greater quality and using fewer personnel.

CENTRA 2000 supports the engineering process from preliminary design through release-to-production. The data management tool enables its user to be fully confident carrying out configuration management, project management and informational management jobs.

Designed as a tool to assist companies with standards and regulatory compliance, CENTRA 2000 also can handle audit histories and provides a means to ensure new information is distributed.

Auto-trol's CENTRA 2000 has the ability to interface to external applications, such as computer-aided design packages like Pro/ENGINEER® (Pro/E) applications and its related modules. Users can check in and check out parts, assemblies, drawings and layouts, all to the benefit of the workflow process.

A CENTRA 2000 satisfied user is NOVA Gas Transmission Ltd. who operates Alberta's 13,300-mile-long, high-pressure pipeline. That stretch of pipeline is one of the world's most technologically advanced gas transportation systems.

Auto-trol's CENTRA 2000 was successfully put in place, succeeding an older NOVA custom drawing file system. The new data handling system will facilitate NOVA's engineering work to manage large quantities of engineering drawings, many of which are considered active at any point in time. Pipeline operators now have quick and easy access to drawings and related text documents. An ongoing expansion program is underway, with CENTRA 2000 able to link disparate databases and applications used in NOVA's engineering and purchasing operations. NOVA's design-to-production cycles are expected to be greatly streamlined.

The CENTRA 2000 product has 30 production sites worldwide, says Susan Floyd, president of Centra 2000, Inc., in Houston.

"The data management system is being used widely in the petrochemical industry to enable companies to adhere to ISO 9000 [an industry standard of quality]. It is also being used in the discrete manufacturing industry to provide configuration management of processes, products and documentation," Floyd says.

™ CENTRA 2000 is a trademark of Centra 2000, Inc.
® Pro/ENGINEER is a registered trademark of Parametric Technology Corporation.

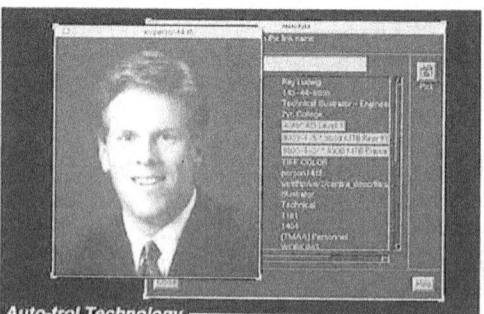

CENTRA 2000 can help manage information on employees, including images.

Stray Light Analysis

Shedding light on a subject may be wanted by some. But in the optics industry, rays of stray light are a troublesome factor.

Based on a Small Business Innovation Research (SBIR) contract from the Jet Propulsion Laboratory (JPL), TracePro™ is state-of-the-art interactive software created by Lambda Research Corporation of Littleton, Massachusetts to detect stray light in optical systems. An image can be ruined by incidental light in an optical system, like static spoils quality sound. To maintain image excellence from an optical system, stray light must be detected and eliminated.

Founded in 1992, Lambda Research Corporation offers stray light analysis, optical system design and analysis, optical testing, process control software and custom software development.

The company's mission is direct: establish a technology bridge between optical and mechanical engineers. That corporate credo is embraced in TracePro, the first optical analysis software program to have as its core, ACIS®, a solid modeling tool that is an industry standard. The ACIS kernel in TracePro provides a vast array of capabilities to create, manipulate, and visualize designs.

Introduced in 1996, TracePro is a Windows® 95/ Windows NT™ application, with roots in the Monte Carlo simulation aspect of ray tracing. This permits the user to compute optical flux as it propagates through a model. Users of the optics software can perform stray light analysis, illumination analysis, and optical systems analysis. The software has an intuitive graphical user interface, familiar Windows pull-down menus, dialog boxes and context-sensitive help.

Lambda Research Corporation president, Edward Freniere, says that TracePro is targeted to optical analysts and mechanical, optical and design engineers across a broad range of industries. "TracePro is the product I have wanted to develop for the past ten years. TracePro is based on the philosophy that there's no reason for an engineer to do the drudgery that a computer can do," Freniere says.

TracePro accounts for absorption, specular reflection and refraction, scattering and aperture diffraction of light. Output from the software consists of spatial irradiance plots and angular radiance plots. Results can also be viewed as contour maps or as ray histories in tabular form.

TracePro is adept at modeling solids such as lenses, baffles, light pipes, integrating spheres, non-imaging concentrators, and complete illumination systems. Optical system designs are capable of being visualized from multiple angles and in rendered views. Objects can be moved and rotated, and modified with optical properties assigned to them.

With Lambda Research's product, the process of going from optical system layout to the completed optomechanical product is streamlined. Tasks that demanded a good part of an hour can be completed in a few computer mouse clicks, saving company time and money.

Marketing of TracePro and other optical analysis software from Lambda Research is now underway throughout the United States, Europe, and the Pacific Rim. The firm's customer base includes corporations such as Lockheed Martin, Samsung Electronics and other manufacturing, optical, aerospace, and educational companies worldwide.

™ TracePro is a trademark of Lambda Research Corporation.

® ACIS is a registered trademark of Lambda Research Corporation.

® Windows is a registered trademark of Microsoft Corporation.

™ Windows NT is a trademark of Microsoft Corporation.

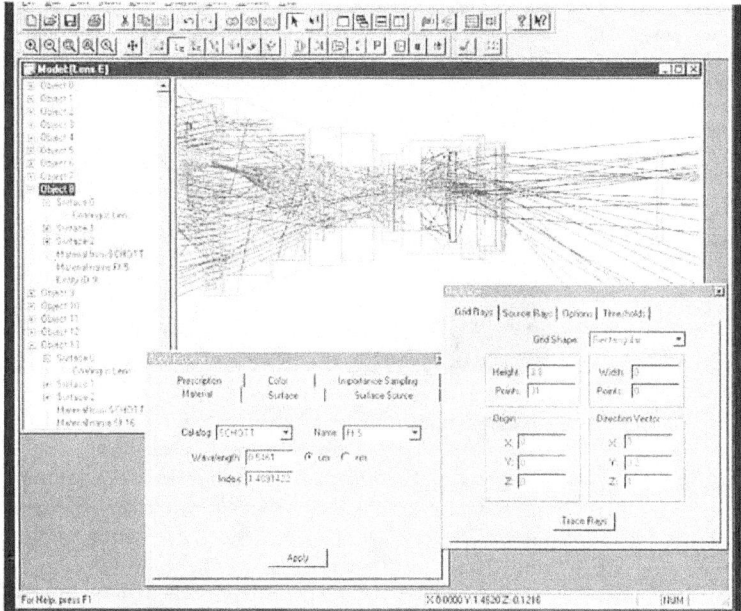

Lambda Research's TracePro software depicts rays of light moving through an optical system. In this instance, the blue light rays indicate the paths of unwanted stray light that can ruin an image. JPL-funded work helped put a new light on this unique software for a range of commercial applications.

Real-time 3D Visualization

Cutting edge 3-D virtual reality technology from NASA used to program and direct planetary explorer vehicles can visualize real-time data in three dimensions.

Fourth Planet, Inc. of Santa Clara, California is a visualization company that specializes in the intuitive visual representation of dynamic, real-time data over the Internet and Intranets.

Founded in 1996, Fourth Planet has as its core a team formerly from the Intelligent Mechanisms Group (IMG) at the Ames Research Center. Over a five-year period, the then NASA researchers performed ten robotic field missions in harsh climes—including underwater in Antarctica, the Kilauea volcano, and Alaska's Mount Spurr—each test designed to mimic the end-to-end operations of automated vehicles trekking across another world under control from Earth. The core software technology used for these missions was the Virtual Environment Vehicle Interface (VEVI), developed at Ames.

VEVI was selected as first runner-up in NASA's 1996 Software of the Year Awards. VEVI is a modular

Based on award-winning NASA telerobotics software, VEVI4 is a powerful tool used to represent complex devices graphically in a 3D environment. Depicted here is the Dante II vehicle during its descent into Mount Spurr, Alaska.

operator interface for direct teleoperation and supervisory control of robotic vehicles. It uses real-time interactive 3D graphics and position/orientation sensors to produce generic vehicle control capabilities. VEVI has been used to control wheeled, legged, air bearing, and underwater vehicles in a variety of environments.

Butler Hine, who was director of the IMG at Ames, and five others partnered to start Fourth Planet. Their calling after leaving NASA is to bring together three important and emerging technologies: Real Time Data, Network Computing, and 3D Graphics and Data Visualization. Fourth Planet contends that the convergence of these technologies will revolutionize the world of computing. It is the mission of the start-up company to drive this revolution with products and services that

put real data, delivered in real time into compelling and useful visual representations.

Not straying too far from Ames, Fourth Planet moved into the NASA/Ames Technology Commercialization Center, a business incubator for start-up companies. Situated in the incubator, Fourth Planet pays a reduced rent and shares administrative cost for copiers, faxes and secretarial assistance. Perhaps more importantly, the group is in residence with 29 other businesses within the incubator, a hub of contacts and knowledge.

Fourth Planet has released VEVI4, the fourth generation of the award-winning VEVI Ames software. VEVI4 is a cutting-edge computer graphics visualization and remote control applications tool, available commercially for the first time. VEVI4 can represent complex devices graphically in a 3D environment.

NetVision is another Fourth Planet product. This package can allow large companies to graphically view and analyze in virtual 3D space such things as the health and performance of their computer network, locate a trouble spot on an electric power grid, or evaluate a company's web of computer links.

Other software products and services are forthcoming from Fourth Planet.

The world of visual computing is undergoing a major shift in technology, explains Fourth Planet literature, "because seeing is understanding." Once confined to universities and high-end research labs with big budgets, the technologies of 3D computer graphics and data visualization are rapidly becoming available to the world at large. In addition, the explosion of the Internet has created an efficient and available infrastructure for the delivery of real-time data.

Combining the technologies of 3D graphics and real-time data collection, Fourth Planet's NetVision creates a virtual representation of large scale, complex networks.

Aircraft Design Software

D esigning new civil air transportation has been made far easier thanks to aircraft design software that inspects, modifies, and integrates the engineering analysis process.

In the 1970s, Ames Research Center initiated an endeavor to greatly improve the conceptual design process. A software tool was developed to give aircraft builders the information they needed to rapidly and accurately forecast during early design phases an aircraft's performance and cost, even the noise it would generate.

Working closely with airframe and engine manufacturers, Ames formulated the AirCraft SYNThesis (ACSYNT) tool. In 1987, Ames and the Virginia Polytechnic Institute (Tech) CAD Laboratory in Blacksburg, Virginia, began to design and code a computer-aided design system for ACSYNT. In 1990, using a Joint Sponsored Research Agreement (JSRA), Ames formed an industry-government-university alliance to improve and foster research and development for the ACSYNT software. This ACSYNT Institute allowed NASA and Virginia Tech to develop and improve computer-aided aircraft design while receiving immediate feedback from industry users of ACSYNT software. American Technology Initiative (AmTech) of Menlo Park, California served as the organizing body for the JSRA, administrating contractual and licensing agreements.

The result? ACSYNT is becoming a predominant software tool for aircraft conceptual design. Moreover, this historical connection with the "customer" provided the foundation for commercializing ACSYNT.

ACSYNT Institute members included NASA's Ames, Lewis and Langley Research Centers and the Naval Weapons Center, as well as major U.S. airframe and engine manufacturers such as Boeing Commercial Airplane, Boeing Helicopter, General Electric Aircraft Engine, Lockheed, McDonnell Douglas, Northrop and Cessna. This alliance lets NASA pool funds and other resources with industry members to support strengthening of the software's capabilities. These enhancements were performed by the space agency and Virginia Tech CAD Laboratory.

From 1987 into 1995, the resulting ACSYNT research and development helped to shape and energize a market for a commercialization effort and the spawning of a new company.

Successful commercialization of ACSYNT has manifested itself in the creation of Phoenix Integration, Inc. of Blacksburg, Virginia. ACSYNT has been exclusively licensed to Phoenix Integration, an outcome of a seven year, $3 million effort to provide unique software technology to a focused design engineering market.

ACSYNT has become a top-notch conceptual level aircraft design tool that provides a 3D computer-aided design (CAD) environment coupled with detailed analysis capabilities. An aircraft concept can be quickly modeled in the CAD interface, and then analyzed using a suite of multidisciplinary modules representing aerodynamic, propulsion, and mission performance parameters. ACSYNT has been successfully applied to high-speed civil transport configurations, subsonic transports, and supersonic fighters.

Phoenix Integration is now marketing, distributing, enhancing and maintaining ACSYNT. The start-up firm has already pioneered Dynamic Integration System™ (DIS) technology, as well as an extensive set of easy-to-use, motif-based tools that interact with any analysis that supports the DIS interface.

It is a Phoenix Integration credo that streamlining the engineering process through more advanced design tools is critical in today's demanding and competitive world. However, recent trends in down-sizing have meant many engineering firms cannot afford to pay engineers to spend time developing software. Phoenix Integration has tackled this dilemma by offering software specifically designed for engineers and at costs far below that if the software was created from inside a company.

Research and development of ACSYNT which, in turn, led to this successful commercialization story, is a blend of innovative technology, a viable market, and leadership to direct the effort.

Using ACSYNT by Phoenix Integration, aircraft designers may enter the dimensions of components and have surfaces created automatically.

™ Dynamic Integration System is a trademark of Phoenix Integration, Inc.

Telemetry Technology

New low-cost telemetry collection products have been made available to the commercial aerospace community by taking advantage of NASA's technology transfer program.

In 1990, Avtec Systems, Inc., located in Fairfax, Virginia, developed its first telemetry boards for Goddard Space Flight Center. Since that time, Avtec has provided innovative, flexible telemetry board and system designs which keep costs low and reliability high. Avtec's customer base has quickly expanded to include a number of NASA centers, major aerospace firms, and a diverse set of commercial and government buyers.

"Over the years, we have worked closely with NASA engineers to design products that meet the needs of the NASA community. We have leveraged this technology to serve the needs of the commercial aerospace community as well," says Avtec's Mary Ellen Orsino, manager of marketing and sales.

Avtec products now include PC/AT, PCI, and VME-based high speed I/O boards and turnkey systems.

By building intelligence into its boards, high data rates can be attained without burdening the host central processing unit.

In addition to Avtec's board-level products, the company offers complete turnkey systems for telemetry acquisition, data quality monitoring, bit error rate testing, and high speed data logging. Along with high speed bidirectional telemetry boards, the company integrates third party products to support several functions, including receiver, bit synchronizer, time code processor, and others as required.

From the beginning, Avtec boards were designed with flexibility in mind by implementing board functionality on Field Programmable Gate Arrays (FPGA). Then, as additional functions were developed, the same board could be used by swapping logic modules. The result? Cost savings to the customer by providing for multiple uses of the same board designs. Additionally, by providing bidirectional capability (both frame synchronization/decommutation and PCM simulation) on a single board, Avtec has matched what competitors charge for unidirectional boards.

The most recent and most successful technology transfer from NASA to Avtec is the Programmable Telemetry Processor (PTP), a personal computer-based, multi-channel telemetry front-end processing system originally developed to support the NASA communications (NASCOM) network. The PTP performs data acquisition, real-time network transfer, and store and forward operations.

Commercial sales of Avtec telemetry hardware was founded on NASA transferred technology.

There are over 100 PTP systems located in NASA facilities and throughout the world. Widespread NASA support of PTP is a result of its low cost and flexible design. Since commercializing the PTP, Avtec has added even more capabilities, created a commercial documentation package, and provides top-notch technical support on the hardware and software for the systems they sell.

"Our NASA customers have responded enthusiastically," says Avtec's Orsino. "We recently delivered over 100 systems to Goddard's NASCOM division for the NASCOM-IP transition, resulting in over $1.5 million in sales for Avtec," she says.

PTPs are scheduled to be used to support Landsat-7, and Stanford University's Gravity Probe B Relativity experiment. "We are also working with many other NASA and commercial aerospace customers, selling PTPs for use all over the world," Orsino adds.

One of Avtec's product lines uses NASA-designed Reed-Solomon Decoder chips.

Avtec has provided telemetry hardware in support of NASA's Communications Network (NASCOM). Shown are NASCOM boards at Goddard Space Flight Center's Simulations Operations Center.

Control Software

Computer software to control automated rovers trekking across the reddish terrain of Mars serve double time in industrial robots that inspect hazardous waste sites, underwater cables, and in the decommissioning of nuclear reactors.

In 1996, Real-Time Innovations, Inc. (RTI) of Sunnyvale, California became the first "graduate" of Ames Research Center's Technology Commercialization Center (ATCC). RTI collaborated with Ames, the Jet Propulsion Laboratory and Stanford University to leverage NASA research to produce ControlShell™ software. RTI's ControlShell is complex, real-time command and control software, capable of processing information and controlling mechanical devices. This capability permits large groups of programmers and projects to share and reuse software objects.

The ControlShell system was used extensively on a cooperative project to enhance the capabilities of a Russian-built Marsokhod rover. As a U.S.-Russian collaborative endeavor, NASA avionics and a ControlShell controller were used in the upgrade.

The Marsokhod rover is being evaluated for eventual flight to Mars. Operated from Earth through electronic links, the planetary tele-rover would depend on control software for day-to-day operations. Additionally, the long delay times of communications between Earth and Mars, upwards of 20 minutes each way, demand a high degree of computer and software support.

RTI's ControlShell empowers its users to build applications from small code objects called components. These components let developers share and reuse code on a scale never before possible. Other key features of ControlShell include its ability to graphically automate both sophisticated feedback-loop control and event-driven reactive logical programming. Another attractive

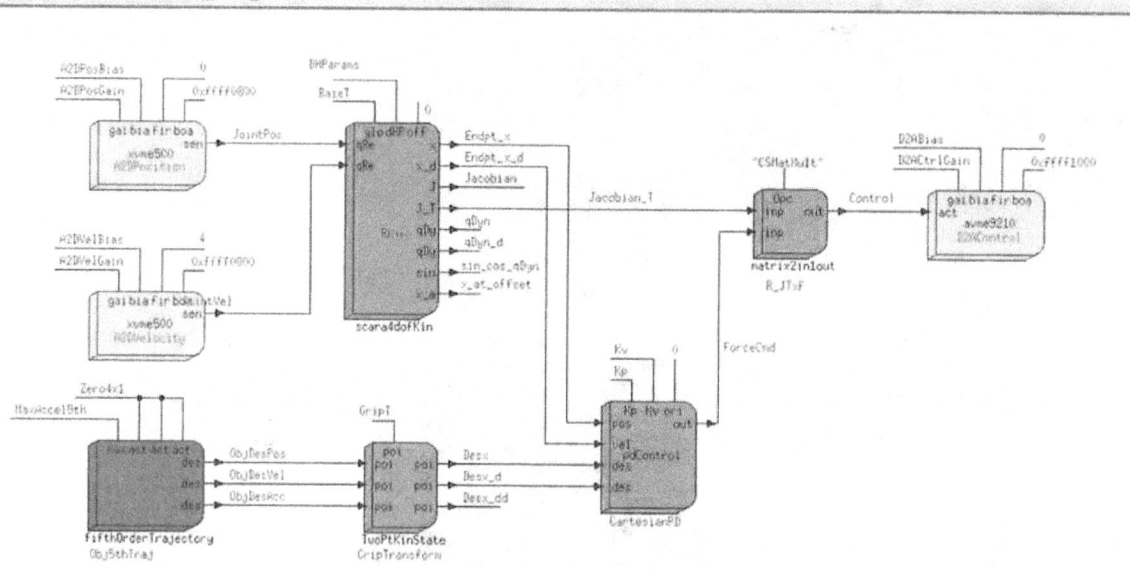

Ames Technology Commercialization Center spurred computer software to run space robots bound for planetary distances to find Earthly applications, such as hazardous waste site inspection.

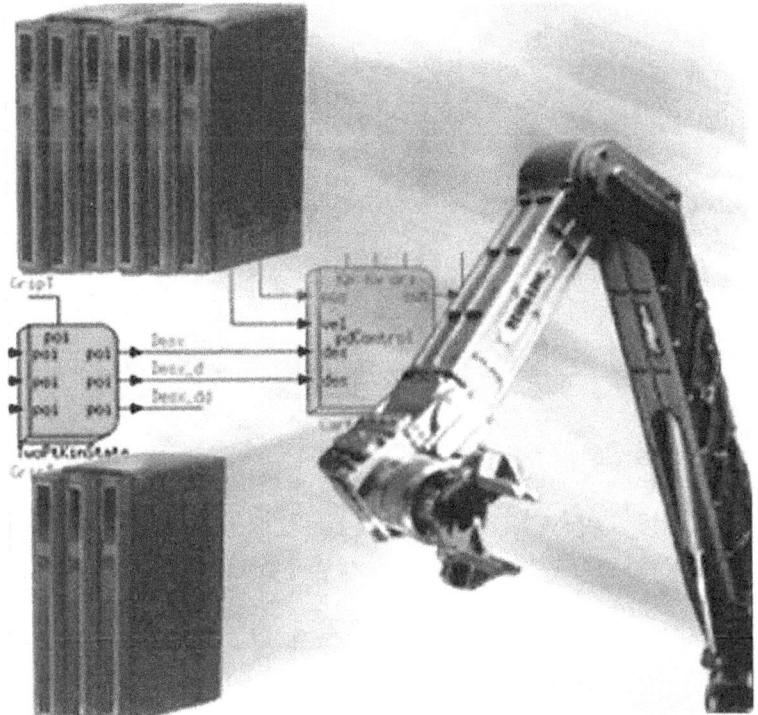

ControlShell allows large groups of programmers and projects to share and reuse software objects. RTI software was the first graduate of the Ames Technology Commercialization Center.

distinction of the software is that novice programmers can easily build structured, quality code. On the other hand, the most talented of programmer can effortlessly extend the framework and build large, complex systems.

Like any job, having the right tool greatly increases productivity. ControlShell offers powerful tools that merge design, translation, execution and analysis. By sharing components, ControlShell increases organizational and inter-site productivity. It also saves time in documentation, verification, and maintenance.

One ControlShell tool is RTI's StethoScope®. As a real-time data collection and display tool, StethoScope allows a user to see how a program is running without changing its execution. A live, graphical window allows the user to monitor variables, archive data, collect time histories, and change values—all without affecting how a system is executing.

RTI has successfully applied its software savvy in other arenas, such as telecommunications, networking, video editing, semiconductor manufacturing, automotive systems, and medical imaging.

Work on the Marsokhod project demonstrated the value of the ATCC incubator program. Taxpayer supported research was leveraged by placing NASA technologies in the entrepreneurial hands of RTI.

"NASA Ames made a huge contribution to our technology," states RTI President and CEO Stan Schneider. "The ATCC catalyzed our efforts to bring this technology to market. These cooperative efforts allow important research results to impact a much wider audience. They are crucial to the efficient use of our national research talent," Schneider says.

™ ControlShell is a trademark of Real-Time Innovations, Inc.

® StethoScope is a registered trademark of Real-Time Innovations, Inc.

Pressure Measurement Sensor

At one time or another, everybody is forced to work under high pressure. But that stress is no match for the cruel and harsh conditions inside internal combustion engines.

FFPI Industries, Inc. of Bryan, Texas is the manufacturer of fiber-optic sensors that furnish accurate pressure measurements in internal combustion chambers. Such an assessment can help reduce pollution emitted by these engines.

A chief component in the sensor owes its seven year-long development to Lewis Research Center funding to embed optical fibers and sensors in metal parts. NASA support to Texas A&M University played a critical role in developing this fiber optic technology and led to the formation of FFPI Industries and the production of fiber sensor products.

FFPI Industries draws its name from use of a fiber optic Fabry-Perot interferometer. This sensing element is the basis for the company's in-cylinder pressure transducer. Engine pressure sensors and signal conditioning units were introduced in 1995 as the first products from FFPI Industries.

Robert Atkins, President and CEO of FFPI Industries predicts that a vast market for the firm's products is likely to emerge. "Sensors are the key to industrial monitoring and control systems designed to lower equipment maintenance cost, improve fuel economy, reduce atmospheric pollution, and provide a safer workplace," Atkins says.

FFPI sensors are enclosed in a housing similar in dimension to a spark plug. A metal jacket protects the device from the high temperatures and pressures that it is to

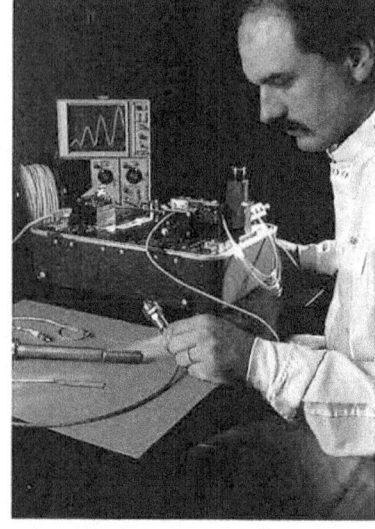

When mounted in an engine, FFPI Industries' fiber optic sensors can relay information to a control system that automatically adjusts the engine for smoother operation.

measure. The sensor can be mounted in a threaded port in the engine's cylinder head or in a standard pressure monitoring port external to the engine.

Fiber optic technology allows the sensor to survive at much higher temperatures than conventional pressure sensors. Additionally, the FFPI sensor is immune to electromagnetic interference, offering another advantage over conventional sensors for industrial applications. Once in place, the sensor relays information to a control system that can automatically adjust the engine for smoother operation.

Clean Air Act provisions and energy savings are sparking keen industrial interest in the FFPI sensor. Legislation has called for operators of large internal combustion engines to seek a reduction in harmful emissions to 10-50 percent of present levels. Retrofitting

the largest class of reciprocating engines with precombustion chambers and/or catalytic converters would entail significant cost. Utilizing the fiber optic engine pressure sensor as one economical solution to emission reduction in large engines is a reasonable idea.

FFPI sensors have measured pressure reliably for thousands of hours in large Clark and Worthington engines in facilities operated by Tenneco Energy Corporation and Transcontinental Gas Pipeline Company. Some 40,000 hours of in-cylinder operation by the sensors, without mechanical failure, have been attained in a pipeline engine. Pressures up to 1000 psi under maximum load at 300 rpm were experienced by the sensors, along with soaring temperatures upwards of 300 degrees centigrade. The sensors have also operated in General Electric and AMD Diesel Locomotive engines under even more demanding conditions: temperatures over 800 degrees Celsius and pressures over 2500 psi.

The simple, rugged design of the sensor offers the potential for mass production at low cost. The company envisions many other sensing applications for the basic technology, including:

- vibration sensors for turbine engines and pumps;
- temperature sensors for semiconductor integrated chip processing;
- biomedical pressure and temperature sensors;
- embedded strain and acoustic sensors for "smart structures" such as aircraft wings and bridges; and
- sensors for monitoring chemical constituents in gases and liquids.

Widespread application of the new technology is foreseen, says Atkins of FFPI Industries, from natural gas transmission, oil refining and electric power generation to rail transport and the petrochemical paper product industry.

Graduate student Han-Sun Choi (left) and Dr. Henry Taylor of Texas A&M University test fiber optic temperature sensors. FFPI Industries tests each of its sensors on an engine with a computer calibration system before shipping.

Filmless Radiography

Liberty Technologies, Inc. of Conshohocken, Pennsylvania, has seen the future of industrial radiography and it is filmless. Thanks to the company's RADView™ technology, licensed from Quantex Corporation of Rockville, Maryland, that future vision is now.

Radiography is implemented in a number of aerospace, oil and gas, and utility applications. It is particularly useful for visualization of cracks, erosion and corrosion problems, and checking the nature of welds, assuring integrity by direct imaging of internal structure. Defects can be discerned in a noninvasive manner. Costly and sometimes deadly catastrophic failures can be prevented through radiography.

RADView is a new filmless technology, a total imaging solution for the conversion of radiographic film records to digital format and digital acquisition of radiographs.

R. Nim Evatt, president and CEO of Liberty Technologies considers the RADView as the first practical advance in technology for industrial radiographic imaging since the discovery of x-rays over 100 years ago.

The system works much the same way as a common medical x-ray—the object to be imaged is first exposed to radiation. But using RADView, a thin and flexible Storage Phosphor Screen—in place of conventional film—records and stores the image. This phosphor screen is a solid-state media. A stored image can be read with a laser and directly digitized for electronic viewing. After viewing or archiving is complete, the screen can be erased, ready for the next image.

Initially prototyped under a Small Business Innovation Research (SBIR) contract between Quantex Corporation and Langley Research Center, the solid-state radiography system can capture latent images with wider dynamic range at lower x-ray or gamma doses than conventional film systems. NASA-sponsored work by Quantex involved the investigation of a technology called Electron Trapping (ET®). That research led to a solid-state film that uses a special class of photoluminescent materials to capture radiographic images.

In June 1994, Quantex signed a license agreement with Liberty Technologies, which further developed and now markets filmless radiography systems for inspection of industrial components. The core work of Liberty is to provide diagnostic, condition monitoring and nondestructive evaluation programs, products and services to manage critical assets and processes in industries worldwide.

RADView converts existing radiographic film to a digital format and digitally acquires images employing the patented phosphor technology. A trio of benefits is immediately realized by the RADView technology: reduction of exposure times and errors; film waste and expense is eliminated; and the efficiency of data management and precise image analysis is boosted.

Liberty's commercial systems rapidly acquire and digitize radiographic images of objects, which then can be viewed and enhanced electronically. The system's software enables a much greater range of information to be gleaned.

Yet another advantage of the RADView approach is that, in general, the solid-state film can be exposed, read, erased, and re-exposed indefinitely until mechanical wear replacement is required. Furthermore, the solid-state film allows electronic archiving and transmission of images. Typically, radiographic films begin to degrade after roughly 10 years. Liberty's digital images, however, can be stored to optical media for up to 100 years and beyond without information loss as well as at low cost.

Liberty has already begun to introduce the RADView to selected overseas markets that require industrial radiography as it is a billion dollar market worldwide.

™ RADView is a trademark of Liberty Technologies, Inc.
® ET is a registered trademark of Quantex Corporation.

RADView provides efficient digital radiography of industrial components to catch wear, erosion, or other defects. The initial research was sponsored by NASA through a Small Business Innovation Research contract.

Super-compact Laser

A super-compact laser for dedicated manufacturing applications was made possible by NASA's need for laser instruments to study Earth as well as investigate the distant shores of Mars.

Microcosm, Inc. of Columbia, Maryland produced the portable FarField-2 laser for field applications that require high power pulsed illumination. The compact design was conceived through research at Goddard Space Flight Center on laser instruments for space missions to carry out geoscience studies of Earth.

An exclusive license to the key NASA patent for the compact laser design was assigned to Microcosm.

Because the FarField laser is so small and can use such a range of power supplies, it is an ideal portable laser for field applications. The commercial utility of the solid-state laser equipment appears promising.

Microcosm's FarField-2 is a miniature solid-state ultraviolet laser, using a Neodymium-Yttrium-Aluminum-Garnet (Nd:YAG) crystal. It also produces green or infrared outputs. Weighing less than 22-pounds, the laser has dimensions of 10.5 x 12 x 4 inches.

FarField-2 has several advantages over excimer lasers, argon-ion lasers, or nitrogen lasers. While compact, the device provides three wavelengths simulta-

The power supply and the FarField-2 laser are packaged together in a small air-cooled housing. Despite its size, the laser is robust making it appropriate for industrial, laboratory and mobile use.

A beam from the FarField-2 Laser is captured in a glass box. Microcosm, Inc. produces lasers for field applications that require high power pulsed illumination.

neously, has low power consumption, does not need water cooling or gas supplies, and produces nearly ideal beam quality.

The power supply and the FarField-2 laser are packaged together in an air-cooled housing. Despite its diminutive dimensions, the laser is robust making it appropriate for industrial, laboratory and mobile use. Microcosm sees applications for the FarField-2 in materials science and processing, as well as taking on duties in biology and medicine. Examples of commercial applications include diamond marking, semiconductor line-cutting, chromosome surgery, and fluorescence microscopy.

The properties of the FarField-2 laser make it effective over long distances. That attribute and others is why NASA developed the technology for laser altimeters that can be toted aboard spacecraft. Pulses from the device can fire many times each second. By measuring the length of time it takes for the light to return to the spacecraft-mounted laser, scientists can determine the distance to a planet's surface. Data collected can be used to construct highly accurate topographic maps.

The FarField-2 has been used in combination with microscopes to create spot sizes of less than one micrometer in diameter. Capable of reaching high energy levels, the laser can create a beam that behaves like a cutting tool. That penetrating power can be utilized for drilling holes and cutting lines, critical applications in the semiconductor industry, in materials science and materials processing.

Dr. Patrick Huddie, CEO and Dr. Wayne Moore, President of Microcosm, Inc., display the FarField-2 Laser. Research on laser instruments for NASA missions made the compact design possible.

Improving Communications Systems

Orbital and flight test long range communications received a high-frequency boost at Dryden Flight Research Center after the NASA center approached Angle Linear of Lomita, California.

The Space Shuttle has many communications systems which are used throughout a typical mission. Voice communication with the Shuttle is usually maintained on S-band and ultra-high frequencies (UHF). While standard GRC-171 radios that operate in the UHF range are extensively used throughout the world, they were not designed for long range communications. In particular, the sensitivity of these UHF radios needed to be enhanced to pick up the weak signals of spaceborne vehicles and distant terrestrial communications.

Given that the radio spectrum has become increasingly congested, the ability to hear extremely weak signals requires greater receiver sensitivity. Receivers must exhibit a high dynamic range and an ability to operate in the presence of many other strong signals without being overwhelmed by interference.

Dryden Flight Research Center was using externally mounted antenna preamplifiers to solve the problem. But, this solution limited the system usage to single frequency STS operations. Dryden realized that a preamplifier mounted inside of the radio could provide the required gain for STS operations while allowing the multimillion dollar tracking communications system to also be used to provide communications for local flight test operations.

Angle Linear is a manufacturer of linear radio frequency products and peripherals for communications. The company specializes in preamplifiers, receiver front end filtering and receiver multi-couplers for a variety of communications applications. In business for some two decades, the group had long supplied research and development and production equipment for business, cellular, aerospace, radio astronomy and amateur needs.

Angle Linear set about the task of solving Dryden's quandary. "This is where Angle Linear provided its unique service," said Michael Yettaw, work leader of the Dryden Communications Facility. "Large manufacturers of preamp products are generally not interested in manufacturing small quantities of custom designed equipment." He also said that companies like Angle Linear that build small quantities of unique equipment to NASA specifications provide a service crucial to research and development.

The solution was a receiving preamplifier specially crafted for NASA and the GRC-171 radio. The results were dramatic, recounts Angle Linear's Chip Angle, general manager and owner of the company.

Substantially better sensitivity than ever before was achieved, and without any interference. Communications with the Space Shuttle is now much more reliable, especially on orbits where the UHF antennas are not in the best of orientation relative to the ground site. NASA Dryden could now carry out reliable communications with the Shuttle on UHF, even from a distance of 3,000 miles as the space plane flew over the Atlantic Ocean. Additionally, Dryden could support local missions without purchasing additional equipment.

The success story has carried over to other NASA projects underway at Dryden, such as the Mir Space Station communications support effort. The preamplifier system is also currently under evaluation by other NASA centers.

For Angle Linear, the NASA fix turned profitable as a very sensitive, high dynamic range amplifier for the commercial communications industry. A preamplifier line was greatly expanded to cover a broader range of frequencies, providing the same sensational improvement to other areas of communications. These include business, government, trucking, land mobile, cellular and broadcast.

Originally developed for NASA Space Shuttle communications, Angle Linear's receiving preamplifier has a very sensitive high dynamic range. The system is used in a variety of communications applications, including business, government, tracking, and cellular.

Carbon Fiber Composites

H yComp®, Inc. of Cleveland, Ohio has recast Lewis Research Center development of PMR-15 polyimide resin into a line of high temperature carbon fiber composite products to solve wear problems in the harsh environment of steel and aluminum mills.

Early NASA work on advanced materials included the need for processable, high-temperature-resistant matrix resins. A polyimide resin proved well-suited for fiber-reinforced advanced composite aircraft engine components, giving these engine parts higher strength and longer wear life.

HyComp was founded on shifting NASA's PMR-15 polyimide resin technology from aerospace and military requirement to the commercial world. The result was WearComp®, self-lubricating composite wear liners and bushings.

A durable and rugged material, WearComp combines carbon graphite fibers with a polyimide binder. The binder, in conjunction with the fibers, provides the slippery surface, one that demands no lubrication, yet wears at a very slow rate. Strength, stiffness, and structural integrity are drawn from WearComp's graphite fibers.

WearComp typically lasts six to ten times longer than aluminum bronze, a time-honored material widely used in mill operations. Unlike bronze, which can gall the mating steel surface, WearComp polishes the same surface and imparts a self-lube film for years of service. It is designed for continuous operation at temperatures of 550 degrees Fahrenheit and can operate under high compressive loads.

HyComp's WearComp self-lubricating bearings negates a mill manager's most aggravating maintenance problems—grease. Grease is expensive to buy and apply, and costlier still to collect and dispose of. Whereas grease on hot mill applications can char and seize moving bearing surfaces, use of WearComp can reduce that problem dramatically. Without grease or sticking problems, worker hours are reduced, downtime of machinery is lessened, and there is a reduction in materials destined for the scrap heap.

A wide variety of chemical agents, including acids, hydraulic fluids, oils and grease, leaves WearComp virtually unscathed. Impact tests have shown the material to be far more damage tolerant than its aluminum bronze counterpart.

WearComp materials help solve downtime problems on flash trimmers for automatic butt welders on continuous steel processing lines. Composite liners made of the material wear so slowly, operators report getting a year or more of service. In comparison, aluminum bronze liners are typically replaced every 13 to 16 weeks. Similarly, guides and rails last five times longer after switching to WearComp liners.

Using computer assisted design techniques, WearComp materials can be formed to specific, dimensionally accurate wear liners for wedges, pockets, and other unique shapes. Rubbing blocks, valve seats, wheels, pulleys, wear shoes and insulators are among the items produced using WearComp.

Because of the material's uncommon properties, bearings made of WearComp can help relieve mill maintenance headaches. Use of the composite material is gaining acceptance among a growing number of mill operators and maintenance personnel.

WearComp materials add months to the life of equipment, making possible major turnarounds in maintenance programs. These long-life, self-lubricating bearing materials are now being used throughout the basic metal industries, from bar and slab casting, through the hot rolling mills.

Nearly all of the aluminum strip processing in the United States and Canada and most steel hot rolling mills and finishing lines are now utilizing these composite materials.

® HyComp and WearComp are registered trademarks of HyComp, Inc.

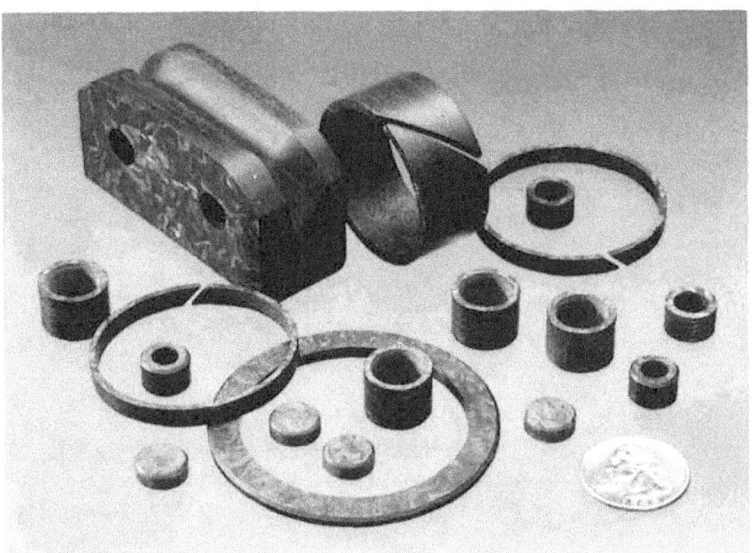

NASA materials research made possible a new class of self-lubricating, high-temperature carbon fiber composites now widely used in aluminum hot mills and steel rolling mills.

Industrial X-ray Imaging

A breed of high-speed camera technology has evolved from a NASA need to study effects of microgravity on science and application experiments carried into space.

For NASA, understanding changes in metal formation and crystal growth while exposed to the space environment—with many responses exceedingly subtle—called for high resolution, high frame rate video technology (HHVT). Studying flame in microgravity, in another instance, as part of combustion research experiments performed on the space shuttle and future space station, dictates use of such tools. In addition, NASA required that HHVT equipment not only had to record information, but process and transmit the data.

In 1990, Lewis Research Center jointly sponsored a conference with the U.S. Air Force Wright Laboratory focused on high speed imaging. The program was organized as a forum where researchers from industry, universities and government could be brought together to discuss the state of knowledge in image gathering, coding, and processing methods.

That conference, and early funding by the Lewis Research Center, helped to spur work by Silicon Mountain Design, Inc. (SMD) of Colorado Springs, Colorado to break the performance barriers of imaging speed, resolution and sensitivity through innovative technology. Continued funding by Wright Laboratory based on this early work led to major breakthroughs in high speed imaging technology.

Later, under a Small Business Innovation Research (SBIR) contract with the Jet Propulsion Laboratory, SMD also designed breakthrough technology which does for high performance imaging what the graphic equalizer does for home stereo equipment. A graphic equalizer in high quality audio, for example, suppresses undesirable hiss while boosting preferred audio frequencies.

Similarly, SMD's "Max-Res" real-time image enhancing camera yields superb, high quality, images in 1/30th of a second while limiting distortion. The result is a rapidly available, enhanced image showing significantly greater detail compared to image processing executed on digital computers.

The company is working with leading medical equipment manufacturers in applying Max-Res technology for radiographic and pathology-based medical applications.

SMD's high speed digital camera work had led to a variety of industrial imaging markets, from airborne spectroscopy to non-invasive inspection of integrated chips and other micromachined parts. Also, Polaroid Corporation is using the technology in a new photon tunneling microscope, while Hewlett Packard is applying the camera to x-ray inspection devices.

SMD work has led to the development of high-speed cameras incorporating state-of-the-art charge-coupled devices (CCD) that produce frame rates ranging from 100 to 10,000,000 frames per second. Wright Laboratory and the Naval Research Laboratory are using the cameras for numerous military applications.

Germany's Qtec is using SMD camera technology to provide automated semiconductor inspection equipment. This industrial vision system, coupled with the requisite computer and software, can allow thousands of component parts in electronic circuitry to be examined per minute.

Silicon Mountain Design's work earned them the 1995 and 1996 Industrial/Manufacturing SBIR Technology of the Year Award. SMD also won the grand prize for the SBIR Technology of the Year in 1996 for a million-frame-per-second camera. Presented by the Technology Utilization Foundation, the award was given to SMD "for developing an innovative new technology through the Small Business Innovation Research program, resulting in the improvement of everyday life and the betterment of mankind."

Silicon Mountain Design's digital camera can produce high quality x-ray images for industrial and medical purposes. NASA's need to detect subtle changes in materials made in microgravity spurred the digital camera's development.

Cleanroom Contaminant Monitor

Helping to keep the work-setting free of airborne molecular contamination is a task of Femtometrics, Inc. of Irvine, California. The company's Real-Time Non-Volatile Residue (NVR) monitor was developed under a Small Business Innovation Research (SBIR) contract for John F. Kennedy Space Center (KSC) in Florida.

Femtometrics had responded to Kennedy Space Center's needs by proposing an advanced, highly sensitive surface acoustic wave (SAW) microsensor capable of detecting sub-monolayer deposition in cleanrooms where aerospace systems are assembled.

Initial criteria established by KSC called for a technology that could regulate the accumulation of nonvolatile residues in cleanroom environments. The smallest speck of material can harm sensitive payloads being prepared for launching. An ultra-clean work environment for assembly and final check-out of a spacecraft is a must.

As specialists in trace analysis and instrumentation, Femtometrics had responded years earlier to an SBIR solicitation by Langley Research Center. Langley's interests centered on size distribution and mass concentration of aerosols and chemical vapor in the stratosphere. NASA SBIR Phase I and Phase II contracts were awarded to Femtometrics resulting in a highly sensitive aerosol detector for environmental researchers.

Stimulated by the SBIR contract wins, the company set about developing proprietary SAW resonator technology. A new type of sensor evolved from the research.

The advanced sensor has the ability to measure a range of chemical vapors by applying chemical-specific coatings on the sensing surface. The SAW proved extremely sensitive—two orders of magnitude greater than a competing, NASA-developed instrument built to conduct the same environmental monitoring function.

Femtometrics equipment can employ multi-SAW sensor array systems. These offer a range of absorption characteristics. When exposed to a chemical-vapor environment, the arrays produce a "fingerprint" of the specific chemical vapor or vapors present. Coupled with custom-programmed adaptive neural network software, the chemical compound fingerprint is analyzed and identified.

William Bowers, President of Femtometrics, explains that a prototype Real-Time NVR monitor was used as an engineering tool during the ground integration and launch processing of the Hubble Space Telescope replacement optics and science instruments during its first servicing mission in 1993. Hubble's second servicing visit by astronauts in 1997, Bowers adds, made use of three passive sensor elements provided by Femtometrics which successfully measured on-orbit contamination.

Commercial applications of the Real-Time NVR Monitor now include Class 1 cleanrooms at semiconductor and hard-disk manufacturing plants. Also, the monitoring system is a process tool to measure condensable contaminates in ultra-high purity operations that use gas and solvents.

Other federal organizations, such as the Department of Defense and the Department of Energy, have found need for the SAW chemical-vapor detection systems built by Femtometrics. Applications include scientific research and chemical warfare detection.

Real-Time Nonvolatile Residue Monitor developed by Femtometrics, Inc. is built to monitor for contaminants in cleanrooms for spacecraft check-out but has found a new home in factories of microelectronic chip makers.

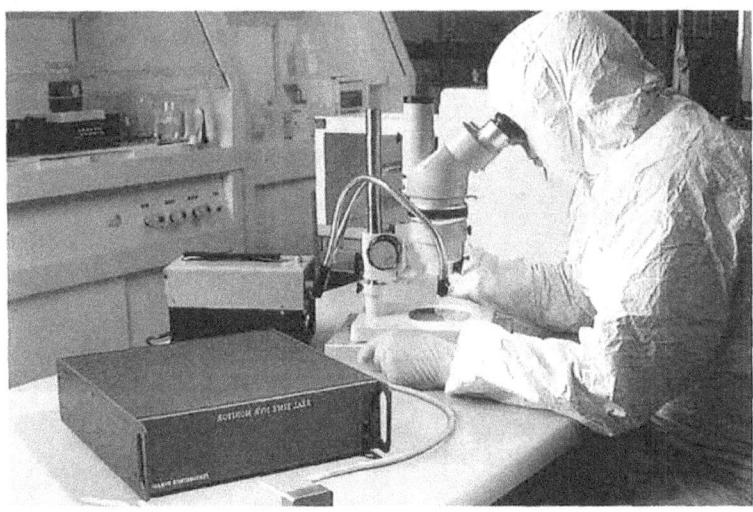

One use of Femtometrics, Inc.'s NVR-200 is in cleanrooms and in process control by semiconductor manufacturers.

NASA's Commercial Technology Network

The NASA Commercial Technology Network (NCTN) extends from coast to coast. For specific information concerning technology transfer and commercialization activities described below, contact the appropriate personnel at the facilities listed or go to the Internet at **http://nctn.hq.nasa.gov.** General inquiries may be forwarded to the National Technology Transfer Center.

To publish your NASA spinoff story, contact the NASA Center for AeroSpace Information or go to the Internet at **http://www.sti.nasa.gov/tto/contributor.html.**

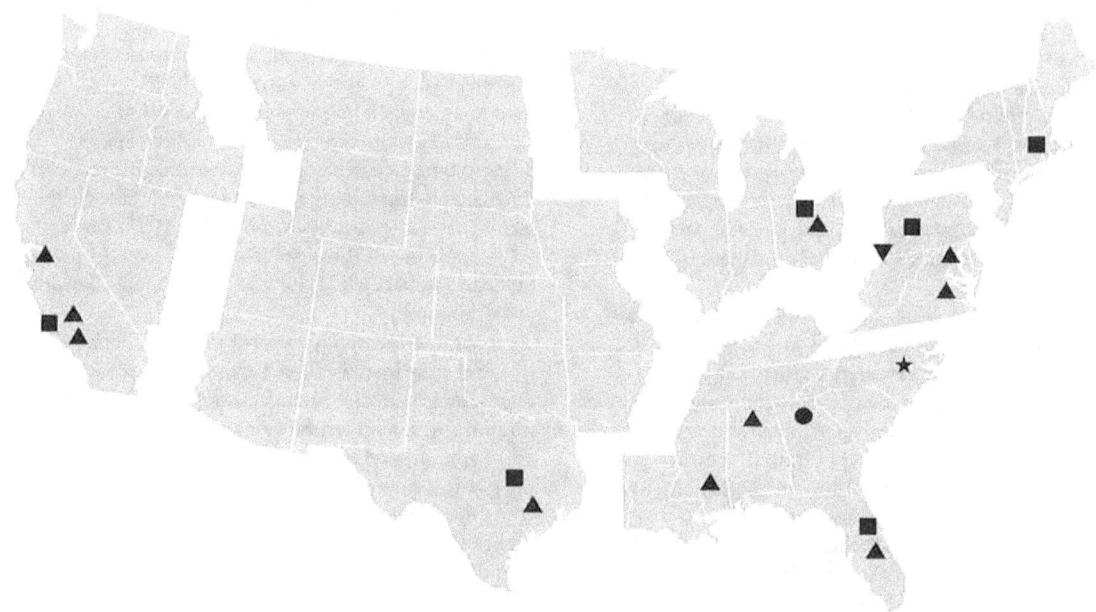

▲ **Field Center Technology Transfer and Commercialization Offices:** represent NASA's technology sources and manage center participation in technology transfer activities.

▼ **National Technology Transfer Center:** national information, referral and commercialization service for NASA and other government laboratories.

■ **Regional Technology Transfer Centers:** provide rapid access to information, technical and commercialization services.

● **The Computer Software Management and Information Center (COSMIC):** offers government-developed computer programs adaptable to secondary use.

★ **Application Team:** provides a range of technology management services including technology assessment, valuation and marketing; market analysis; intellectual property audits; commercialization planning; and the development of partnerships.

▲ FIELD CENTERS

Ames Research Center
National Aeronautics and
Space Administration
Moffett Field, California 94035
Chief, Commercial Technology Office:
Bruce Webbon, Ph.D.
Phone: (415) 604-6646
email: BWebbon@mail.arc.nasa.gov

Goddard Space Flight Center
National Aeronautics and
Space Administration
Greenbelt, Maryland 20771
Technology Transfer Officer:
George Alcorn, Ph.D.
Phone: (301) 286-5810
email: galcorn@pop700.gsfc.nasa.gov

Lyndon B. Johnson Space Center
National Aeronautics and
Space Administration
Houston, Texas 77058
Director, Technology Transfer and
Commercialization Office:
Henry Davis
Phone: (281) 483-0474
email: henry.l.davis@jsc.nasa.gov

John F. Kennedy Space Center
National Aeronautics and
Space Administration
Kennedy Space Center, Florida 32899
Technology Transfer Officer:
James A. Aliberti
Phone: (407) 867-6224
email: jim.aliberti-1@kmail.ksc.nasa.gov

Langley Research Center
National Aeronautics and
Space Administration
Hampton, Virginia 23681-0001
Director, Technology Applications Group:
Joseph S. Heyman, Ph.D.
Phone: (757) 864-6005
email: j.s.heyman@larc.nasa.gov

Lewis Research Center
National Aeronautics and
Space Administration
21000 Brookpark Road
Cleveland, OH 44135
Director, External Programs:
John Hairston, Jr.
Phone: (216) 433-5568
email: john.m.hairston@lerc.nasa.gov

George C. Marshall Space Flight Center
National Aeronautics and
Space Administration
Marshall Space Flight Center, Alabama 35812
Manager, Technology Transfer Office:
Sally A. Little
Phone: (205) 544-4266
email: sally.little@msfc.nasa.gov

Jet Propulsion Laboratory
4800 Oak Grove Drive
Pasadena, California 91109
Manager, Commercial Technology Program:
Merle McKenzie
Phone: (818) 354-2577
email: merle.mckenzie@jpl.nasa.gov

NASA Management Office—JPL
4800 Oak Grove Drive
Pasadena, California 91109
Technology Commercialization Officer:
Arif Husain
Phone: (818) 354-4862
email: ahusain@nmo.jpl.nasa.gov

John C. Stennis Space Center
National Aeronautics and
Space Administration
Stennis Space Center, MS 39529
Technology Transfer Officer:
Kirk V. Sharp
Phone: (601) 688-1914
email: kirk.sharp@ssc.nasa.gov

Dryden Flight Research Center
National Aeronautics and
Space Administration
Post Office Box 273
Edwards, California 93523-0273
Chief, Technology Commercialization Office:
Lee Duke
Phone: (805) 258-3802
email: duke@louie.dfrc.nasa.gov

■ REGIONAL TECHNOLOGY TRANSFER CENTERS

1-800-472-6785 You will be connected to the
RTTC in your geographical region.

Far-West
Technology Transfer Center
University of Southern California
3716 South Hope Street, Suite 200
Los Angeles, California 90007-4344
Kenneth E. Dozier, Jr., director
Phone: (213) 743-2353
email: kdozier@bcf.usc.edu

Northeast

Center for Technology Commercialization, Inc.
1400 Computer Drive
Westborough, Massachusetts 01581
William Gasko, Ph.D., director
Phone: (508) 870-0042
email: wgasko@ctc.org

Mid-West

Great Lakes Industrial Technology Center
25000 Great Northern Corp. Ctr., Suite 260
Cleveland, Ohio 44070-5320
Christopher Coburn, executive director
Phone: (216) 734-0094
email: coburnc@battelle.org

Southeast

Southern Technology Application Center
University of Florida
College of Engineering
Box 24
One Progress Boulevard
Alachua, Florida 32615-9987
J. Ronald Thornton, director
Phone: (904) 462-3913
email: jrthorn@nervm.nerdc.ufl.edu

Mid-Continent

Texas Engineering Extension Service
Texas A&M University System
301 Tarrow Street
College Station, Texas 77843-8000
Gary Sera, director
Phone: (409) 845-8762
email: ecsera@teexnet.tamu.edu

Mid-Atlantic

University of Pittsburgh
3400 Forbes Avenue, 5th Floor
Pittsburgh, Pennsylvania 15260
Lani Hummel, director
Phone: (412) 383-2500
email: lhummel@mtac.pitt.edu

● COMPUTER SOFTWARE MANAGEMENT AND INFORMATION CENTER

COSMIC
382 E. Broad Street
University of Georgia
Athens, Georgia 30602
Tim Peacock, director
Phone: (706) 542-3265
email: timp@cosmic.uga.edu

★ TECHNOLOGY APPLICATION TEAM

Research Triangle Institute
Post Office Box 12194
Research Triangle Park,
North Carolina 27709
Doris Rouse, Ph.D., director
Phone: (919) 541-6980
email: rouse@rti.org

▼ NATIONAL TECHNOLOGY TRANSFER CENTER

Wheeling Jesuit University
Wheeling, West Virginia 26003
David Moran, Ph.D., president
Phone: (304) 243-2462
email: dmoran@nttc.edu

NASA CENTER FOR AEROSPACE INFORMATION

Spinoff Project Office
800 Elkridge Landing Road
Linthicum Heights, Maryland 21090-2934
Walter Heiland, manager
Phone: (301) 621-0241
email: wheiland@sti.nasa.gov

Spinoff Team

Project Manager:
Walter Heiland

Editor/Writer:
Jennifer Munro

Editor:
Amy Marselas

Editorial Assistant:
Lenora Parris

Graphic Design:
John Jones

Photography:
Kevin Wilson

Writer:
Leonard David